SOLDIER-SCHOLARS:
HIGHER EDUCATION IN THE AEF, 1917-1919

1. Colonel Ira L. Reeves, Commandant and President, AEF University, Beaune, France.

HELL, YES!

"Private Williams, take the floor;
How much, please, is two plus four?
Three times seven, minus eight,
Leave just what, you'll kindly state?
If you have six porcupines,
Five fell down and cracked their spines,
Just how many would remain?
Write it down and make it plain."

Now ain't that a healthy way
For a soldier man to play
Ev'ry evenin' more or less?
HELL, YES!--HELL, YES!

"Private Johnson, take the floor;
Please bound Lower Labrador.
How much cheese does Spain import
From the Duchy of Connaught?
If you went to Timbuctoo
What canals would you pass through?
Where's the biggest swamp in Maine?
Tell us that and make it plain."

Now ain't that a lovely song
For a man who's big and strong
'N' aching for some happiness?
HELL, YES!--HELL, YES!

"Private Murray, take the floor;
Who discovered Singapore?
How did Alsace come to France?
Name King William's maiden aunts.
Tell us how Napoleon
Won the battle of Bull Run.
Who in Louis Quinze's reign
First used soup? Now please explain."

Now ain't that a gorgeous tune
For a soldier guy to croon
Night--an' mornin's, too, I guess?
HELL, YES!--HELL, YES!

SOLDIER-SCHOLARS

"Private Perkins, take the floor;
Scan this philosophic law,
Who was Kant and who was Locke?
Why did Hick'ry Dick'ry Dock
Run about and play when he
Might have read philosophy
And learned to talk in high-brow strain?
I dare, you, sir, to make it plain."

Now ain't that a scrumptious way
For a hulkin' man to play?
Next they'll teach us how to dress.
HELL, YES!--HELL, YES!

   __T. G. Brown, Cpl., 51st Pioneer Infantry,
    The Stars and Stripes, March 14, 1919

SOLDIER-SCHOLARS:
HIGHER EDUCATION IN THE AEF, 1917-1919

Alfred Emile Cornebise

AMERICAN PHILOSOPHICAL SOCIETY
Independence Square • Philadelphia
1997

*Memoirs
of the
American Philosophical Society
Held at Philadelphia
For Promoting Useful Knowledge
Volume 221*

Copyright © 1997 by the American Philosophical Society for its Memoirs series. All rights reserved. Cover *illustration*: Outdoor Class in French, University of Grenoble. Still Picture Branch, National Archives, 111-SC-160999.

ISBN:0-87169-221-X   Library of Congress Card No:96-84045
US ISSN: 0065-9738

Table of Contents

Poem: Hell, Yes! iii
Table of Contents vii
List of Illustrations. viii
Dedication Page ix
Map x
Acknowledgments xi
Preface xii
Chapter 1: Introduction 1
Chapter 2: Founding the University 14
Chapter 3: Setting up the Traditional Academic Colleges. 50
Chapter 4: The Maturing University 71
Chapter 5: The Farm School 82
Chapter 6: The Art School 97
Chapter 7: Relations with the French 123
Chapter 8: Other Universities 130
Chapter 9: Last Days 166
Chapter 10: Conclusions 179
Index 193

List of Illustrations

(All are from Still Pictures Branch, The National Archives, Collection 111-SC, and are used by permission).

Frontispiece. Colonel Ira L. Reeves, Commandant and
    President, AEF University, Beaune, France ii
2. General Robert I. Rees, Head of AEF Education 16
3. Dr. John Erskine, Columbia University 20
4. Auditorium under Construction, Beaune................ 52
5. Campus in the Mud and Pershing Athletic Field, Beaune 52
6. New Construction, AEF University, Beaune 53
7. Registration Lines, Beaune 53
8. Gas Engine Class 75
9. Art Training Center, Bellevue 100
10. Doughboy Art Students Drawing from Casts, Bellevue 100
11. Students's Drawings in Architecture, Bellevue 101
12. American Students at Work, Académie, Julian, Paris 103
13. French Reception, Grenoble Chamber of Commerce 103
14. The Sorbonne, Paris 105
15. University of Montpellier 139
16. Trinity College, Dublin, and Doughboy Students 143
17. Soldier-Student Detachment, University of Bristol 143
18. Student Council, University of Lyon Detachment 144
19. Track Team, University of Lyon Detachment 144
20. Dissecting Room, University of Lyon 145
21. Outdoor Class in French, University of Grenoble 145
22. Dining Hall, University of Toulouse 146
23. Lecture Hall, University of Dijon 154
24. Industrial Lab, University of Dijon................... 154

FOR JAN

SOLDIER-SCHOLARS

Sites of AEF Higher Education in France, 1917-1919.

Acknowledgments

Many people have come to my assistance in the preparation of this account. Chief among these are Richard J. Sommers, John J. Slonaker and David A. Keough at the U.S. Army Military History Institute, Carlisle Barracks, Pennsylvania. At the National Archives in Washington, Richard F. Cox and Mitchell Yockelson, both of the Military Reference Division, were, as always, most helpful. Also, the photographs used herein are by courtesy of the Still Pictures Branch of the National Archives. Linda Wheeler, at the Hoover Institution on War, Revolution and Peace, in Palo Alto, California, expeditiously sent me badly-needed copies of the student paper of the AEF University. Lois Leffler, at the Interlibrary Loan Office of the Michener Library, University of Northern Colorado, provided me with a steady flow of pertinent materials. In Beaune, France, the curator of the municipal archives, Elaine Lochot, provided me with many documents. In addition, my cousin, Raymond Guérin and his wife, Lucette, of Héry, France, obtained other materials, documents and photographs in Beaune which greatly assisted me in completing this study.

Finally, at the American Philosophical Society, I wish to acknowledge the unfailing encouragement and friendship of Carole N. LeFaivre-Rochester over the years, as well as the fine editorial work of Susan Babbitt in the preparation of this manuscript.

Preface

THE CONVENTIONAL picture of the doughboy of the American Expeditionary Forces [AEF] in World War I is that of a khaki-clad soldier with the "tin hat"—the "Charlie Schwab bowler"—slightly askew on his head.[1] With gas mask bag hanging at his chest, his bayoneted rifle at high port, he is storming "over the top," to get at the Hun, and gain the expected victory. But at least some American soldiers, after the end of hostilities, were uniformed in other garments, such as—most unexpectedly—artists' smocks, as they rather primly and earnestly sketched nudes in life study classes in any of several art programs they were enrolled in. Others attended classes at the Sorbonne, took medical courses at London's Fellowship of Medicine, read law at the Inns of Court, enrolled in veterinary classes at the University of Edinburgh; and studied French culture and language at Poitiers, Montpellier, and Dijon, and at numerous other French universities and institutes. About ten thousand men were involved in these programs. Ten thousand additional soldier- students attended the AEF's own university at Beaune, 170 miles southeast of Paris in the Côte d'Or, near Dijon, created from nothing in a few short weeks to give yet other aspiring doughboy- scholars academic opportunities at the college level. This fully- fledged school had thirteen colleges, including a medical school and a law school; a thirty-thousand-volume library; a full complement of deans, administrators, and faculty; an instantly-created "college spirit"; a burgeoning athletic program; a band; its own substantial art institute near Paris; and a self-contained farm school at Allerey, near Beaune. For a brief few months in the spring of 1919, the university was the largest in the English-speaking world. Those who attended were expected to maintain military discipline yet balance it with the attributes and ambience of a liberal, open-minded, American-style campus. Beyond the university level, other educational opportunities of various sorts were made available to virtually every soldier in the AEF. Organized by the YMCA and army officials to help in keeping the two million men in Europe occupied while awaiting shipment home, from the start these educational ventures were recognized by many interested men as being much more. There were those both in and out of the army who perceived the military establishment as one of the engines of Progressivism. Even beyond the specific educational programs created at the university level, the army itself was regarded as a "poor man's university," or better stated, perhaps, a university of the people, with influence extending beyond the rank and file. Therefore, though growing primarily out of immediate needs, the AEF's ventures in education at all levels, including college and university, gave substance to the view of the army as the "university in khaki." Not only had this organization apparently made the world safe for democracy, but it now intended to intervene decisively in the lives of its essentially civilian-soldiers to help give substance to an enhanced quality of life in the now saved world. Some even wanted to see the War Department transformed into a National Defense Department, which would combine military training with formal educational opportunities such as those then being pioneered by the AEF, thereby consolidating those efforts and extending them into the future with no doubt salutary results.

But in the short term, in the 1920s and 1930s, the army, because of the rapid development of modern warfare, only enlarged its already well-developed training programs and system of professional military schools, rather than significantly pursuing the liberal educational innovations that it had created in 1919. Even here, there was much lagging behind, though foundations were laid for the rapid acceleration of professional training and education needed with the advent of World War II.

In the meantime, the nation did do something for the discharged soldier, if not primarily to further his education. Since the Revolutionary War, the American people had supported various kinds of pensions for their fighting men. But during World War I, veterans demanded much more, and the U.S. government responded with additional programs. Some were elaborations on precedents, but several additions were made. New allotments and allowances for the support of dependents during the veterans' period of active service were created. There were provisions for hospital and medical care for all veterans, regardless of the nature or origin of their infirmities. Term insurance was provided for members of the armed forces; it could be converted to standard insurance plans within a limited period after the end of hostilities. Vocational rehabilitation for disabled veterans was undertaken to a substantial degree. Guardianship services were created for incompetent and minor beneficiaries under certain conditions.[2] In addition, the scheme of veterans' preference in obtaining civil service posts, which had been practiced throughout the history of the republic, was enhanced. Also, the United States Employment Service, which even sent representatives to France after the armistice, assisted the men in finding employment, and many states enacted legislation to help provide jobs for returning service

PREFACE

personnel.[3] Pensions were extended, especially to disabled veterans. Further gains were made by the passage of a modest, though controversial, bonus bill in 1924, which survived a veto by President Calvin Coolidge. These advancements owed much to the lobbying of the powerful American Legion, founded in 1919, supported by the Veterans of Foreign Wars (VFW) and the Disabled American Veterans (DAV).[4]

In other ways, the veteran received modest compensation. He was granted a $60.00 bonus at the time of his honorable discharge from the service, and was paid travel to his home from point of discharge at the rate of five cents per mile. He was allowed to keep his uniform, consisting of a hat or more likely an overseas cap, an olive drab shirt, woolen blouse and ornaments, one pair of woolen breeches, one pair of shoes, one pair of the cordially-hated and much-cursed leggins, a waist belt, an overcoat, two suits of underclothing, four pairs of socks, one pair of gloves, a toilet set, his barracks bag, and as souvenirs, his gas mask and helmet. He was issued a scarlet discharge chevron to be attached midway between shoulder and elbow on the left sleeve of shirts, blouses, and overcoats, the World War I equivalent of the famous "ruptured duck" of World War II fame. In addition, he was issued a lapel button of bronze or silver, the latter for those wounded in action, to wear on his civilian suit. Victory medals were also struck and appropriate ribbons issued.[5]

Far different would be the lot of the doughboy's successor in World War II. Most notably, those then discharged from military service were the beneficiaries of a substantial G.I. Bill, with its elaborate educational provisions. It is difficult to quarrel with the novelist James Michener, who has asserted that the G.I. Bill, enacted by Congress late in World War II, was one of the most beneficial acts ever passed by that august body.[6] However, the bill's educational benefits rested on precedents that the AEF established late in World War I and in the months that followed.

At a time following the ending of the Cold War, when the armed forces are seeking a reorientation, with some emphasis on educational innovations, it is of interest to examine a time in the army's past when it underwent a similar exercise. This study, in addition, reveals aspects of what life was like in the AEF in a time that seemed so agonizingly long for so many: the weeks and months between the armistice and the sailing for home.

Alfred Emile Cornebise
Greeley, Colorado
May 1995

SOLDIER-SCHOLARS

NOTES

1. The steel helmet, the so-called tin hat, was often styled a "Charlie Schwab bowler," in reference to Charles Michael Schwab (1862-1939) who was president of U.S. Steel, 1901-1903, and of Carnegie Steel, 1905-1916, and chairman of its board from 1916 to his death.

2. For the development of veterans' benefits, see: Richard Severo and Milford Lewis, *The Wages of War. When America's Soldiers Came Home—From Valley Forge to Vietnam* (New York: Simon and Schuster, 1989); Davis R. B. Ross, *Preparing For Ulysses. Politics And Veterans During World War II* (New York: Columbia University Press, 1969); and William Pyrle Dillingham, *Federal Aid To Veterans, 1917-1941* (Gainesville, Fla.: University of Florida Press, 1952).

3. See articles in *Stars and Stripes*, April 4 and 18, 1919.

4. The VFW was founded in 1899, the DAV in 1920.

5. The *Stars and Stripes* published numerous articles on the uniform, medals, and ribbons. See issues of February 21, March 21, March 28, April 4, April 18, May 16 and May 23. The men were advised that Congress had granted them the privilege of wearing their uniforms after discharge, in exchange for which they must "be particular about their conduct, appearance, association and habits." However, the men need not salute or otherwise conform with military discipline. Also on the day of discharge, each soldier was given a three-months' membership in the YMCA association nearest his home. The Y hoped in this way to help serve as "a connecting link between the camp life of the soldier and the program of reconstruction." See discussion in William Howard Taft, *Service With Fighting Men. An Account of the Work of the American Young Men's Christian Associations in the World War*, 2 vols. (New York: Association Press, 1922), 1: 400

6. See James A. Michener, *The World is My Home: A Memoir* (New York: Random House, 1992), 263. Michener notes that the two laudable acts that Congress was responsible for which "helped to improve the quality of our national life," were the Homestead Act and the related Morrill Act, and the G.I. Bill. Regarding the latter, he said: "I judge it to have been one of the best expenditures of public money made in my lifetime, for it helped an entire generation of bright young people improve themselves and make an effort to accomplish something meaningful. The burst of achievements in all fields that the United States saw in the decades following the end of World War II stemmed in large part from the flood of energy released by the G.I. Bill."

CHAPTER 1:
Introduction

ONE OBSERVER in France in late 1918 noted, regarding the American forces engaged there, that "on the morning of November 11, 1918, the battle-cry of the AEF shifted from 'When do we eat?' to 'When do we go home?'" Indeed, the more than two million American servicemen in Europe demanded that they be returned to the United States immediately. As one soldier newspaper recorded it, "[A] fellow informs us that he wants to go home 'toot sweet and the tooter the sweeter.'"[1] In this, they were loudly supported by the home folks. However, this movement was impossible for several reasons. Most important, shipping was simply not available. In addition, the armistice was not a treaty of peace, but merely a cease-fire; hostilities could resume in short order. Moreover, it was recognized that the hand of the government would be strengthened in the peace talks, soon to begin in Paris, by the retention of a substantial military force in Europe. Finally, a large U.S. contingent was needed to occupy the zone in Germany allocated to the Americans, a portion of the Rhineland centering on the Rhine port city of Coblenz.[2]

Headquarters of the American Expeditionary Forces at Chaumont, France, initially responded to these post-armistice conditions in a time-honored manner: spit and polish, frequently held in abeyance during hostilities, reappeared with a vengeance. General Orders No. 207 required that for five hours per day, five days a week, drill and training were to be vigorously pursued, while Saturday mornings would be utilized for "a rigorous inspection of all troops." Soon even veterans of combat found themselves engaging in seemingly endless practice of the things that they had so recently done in earnest, often carried out in cold, rainy weather. Accordingly, many of the men were promptly "bored to tears," and became "soured in mind and heart." Consequently, "they were embittered in a way they never forgot," as Pershing's aide, Colonel George Catlett Marshall, Jr., admitted.[3] Soon problems of discipline and a seriously sagging morale were apparent; multitudes of men simply went AWOL, more often than not bound for Paris, which they now saw no reason not to see. In these conditions, American-French relations suffered, and tension sometimes produced serious clashes, causing Chaumont further headaches.

SOLDIER-SCHOLARS

However, there were other plans and developments, many long in the making, which anticipated this state of affairs and which now rather belatedly emerged. In the first place, several hundred thousand of the doughboys were sent into Germany, fully occupying them for some time to come. In addition, General Orders No. 241, GHQ, AEF, of December 29, 1918, stipulated that in order to create better morale, every effort would be made to encourage athletic and entertainment programs. This order also reduced the hours of drill and military training.[4] Beyond this, leaves and an extensive educational scheme were being considered and would shortly surface. These programs were based on extensive planning accomplished even in the midst of hostilities. While key military personnel were involved, the agency most responsible for the establishment of the contemplated initiatives was the Young Men's Christian Association, the YMCA.

The Y, as it was popularly called, was founded in England in 1844. It arrived in the United States in 1851. In 1856, it began its work with the military in Portsmouth, Virginia, when it received permission to place library books aboard a naval training vessel. During the Civil War, YMCA personnel assisted soldiers in the field, both North and South. After the war, the YMCA expanded its military service by establishing recreational and counseling services for soldiers and sailors. During the Spanish-American War, Y personnel were with the troops in Cuba, Puerto Rico, and the Philippines. In 1898, the Y established an Army and Navy Committee, consolidating its presence among the military forces. Soon, a permanent building appeared at the Brooklyn Navy Yard, followed by similar establishments elsewhere. Afterwards, the Y was a familiar presence on army and navy bases around the world. Subsequently, the Y followed national guardsmen to the U.S.-Mexican border as early as 1911, when tension developed between those two nations, setting up canteen buildings and in other ways aiding the U.S. effort. In 1916, when the trouble intensified, the organization's involvement was even more pronounced. The American commander on the border, General John J. "Black Jack" Pershing, long an admirer of the YMCA, asked that welfare personnel follow the punitive expedition into Mexico, his "having come to expect them to be as much a part of army equipment as the army mule or the commissary cook."[5] Therefore, by the time of the declaration of war by the United States on April 6, 1917, the Y was a familiar sight on the military scene. As a matter of course, it accompanied the AEF to Europe and was soon active in creating an extensive network of welfare activities.[6] In addition, it undertook detailed planning in anticipation of further welfare requirements within the AEF when peace came. The armistice caught even the Y by surprise, however, and it moved expeditiously to meet the vast needs so suddenly apparent.

One of the most impressive of the programs that the Y contributed to preparing was that of granting leaves. General Pershing had hoped that all men of the AEF could see Paris, but as he observed in his final report to the War Department, "The crowded condition of the city during the Peace Conference, transportation difficulties, and other reasons, made it necessary to limit the number of such leaves."[7] Therefore, the ensuing provisions for seven-day furloughs, which included accommodations and food, that were soon instituted, emphasized other areas. Leave centers had first been established by the YMCA early in 1918, centering on the French resort of Aix-les-Bains, in the French Alps. With the coming of the new conditions, the YMCA created a Leave Area Department and prepared to meet the provisions of General Orders No. 14, of January 18, 1919, which set forth details for the program, already provided for in earlier General Orders. Eventually, there were nineteen centers, located in the following areas: along the Brittany coast; in the French Riviera area; focusing around Clermont-Ferrand, in France's central massif in the region of Auvergne; in the French Alps, centering on Chamonix; in Provence; and at Biarritz, on the Atlantic coast. By mid-February 1919, it was necessary to schedule special leave trains, seventy-eight of which were eventually servicing the nineteen leave areas.[8] Despite their disappointment at missing Paris, the doughboys were soon enthusiastic about the leave centers and the opportunities to revel in luxury, and to participate in the extensive entertainment and recreational programs in what were, after all, highly desirable vacation spots. By May 1919, at which time almost a million of the two million men in Europe had returned to the United States, nearly 450,000 soldiers had taken advantage of leave opportunities.[9]

If not all of the AEF's doughboys went on furlough, almost all of them participated in the extensive athletic program designed to saturate the army with a variety of sports. As the *Stars and Stripes* asserted, "Probably no other single characteristic of the Yanks will prove so far reaching in its effect upon European minds as their love of athletics."[10] The Y had been involved with athletics before the armistice, but because of the wartime conditions, no concerted, systematic program was possible. General Orders No. 241, of December 29, 1918, transformed the AEF into "a vertible [sic] beehive of athletic industry."[11] The men in charge were Elwood S. Brown, chief athletic director of the Y, and Colonel Wait C. Johnson, the AEF's chief athletic officer. Brown had long been associated with the Y's athletic programs, notably in the Philippines, where, in 1910, he had established a series of international sporting events known as the Far Eastern Games.[12]

The AEF athletic program was fourfold. First, it featured mass games for all AEF personnel, emphasizing "athletics for everybody." In this connection, one observer has concluded that the system of mass athletics, whereby "every man [was] in the game," ultimately "became the YMCA's chief contribution to the physical welfare of the Army in the United States."[13] Other parts of the program featured athletic pageants and demonstrations held for the benefit and enlightenment of Allied nations; official AEF championships culminating in a great finals in Paris, May 21-31; and an Inter-Allied athletic contest, open only to soldiers of the Allied armies—a quasi-military Olympic Games held in Paris.[14] These were intended to be "a fitting close to the greatest military struggle of modern times." The United States welfare agencies and the U.S. Army paid all expenses, which proved to be considerable, as it was soon decided that a new stadium was necessary. The U.S. Army, the French government and the Y cooperated in its construction. Located near Paris, and named Pershing Stadium, it could seat thirty thousand spectators. The games, held June 22 to July 6, 1919, were a great success.[15]

General Orders No. 241's provisions also applied to entertainment, and AEF Entertainment Bulletin No. 1 of January 28, 1919, elaborated. Consequently, almost as comprehensive as the organization of athletics were the provisions for "suitable" entertainment programs, which, to the extent possible, were to be provided each night in every important center occupied by American troops. At this time, Colonel John R. Kelly, of G-1 [Administration], GHQ, at Chaumont, was put in charge of all AEF entertainment.[16] As in other matters, the Y had long been involved, it being clear to the organization's leadership that "entertainment [was] not a luxury but a necessity—as vital as sugar to food."[17] Previously, in October 1917, the Y had set up an office in New York to devise programs to meet the entertainment needs of the AEF. Movies, being readily available and easily presented, had long been a staple of AEF amusement. In the summer of 1917, General Pershing requested that the organization take charge of films. It responded by forming a Cinema Department. Another of its agencies was the Over There Theater League, created in April 1918, for sending American entertainers overseas. Of the 828 professional performers who subsequently went

INTRODUCTION

overseas to entertain the AEF, 454 were recruited by the League. American entertainers were augmented by some five hundred French professionals, mainly working in the leave centers.[18]

In addition to these thespians, the Y also sent over two hundred expert song leaders to train Y secretaries in song leading. The organization also distributed millions of copies of a small book, *Popular Songs of the AEF*, which served as the basis for the mass singing programs that were so popular in the U.S. Army during World War I.[19] After the armistice, entertainment activities rapidly expanded. One of the most extensive was the development of soldier shows. There was plenty of talent in the AEF; the problem was to assemble it. An innovative means of accomplishing this was the so-called play factories. The largest was at Tours, with others at Paris, Bordeaux, and St. Nazaire. In these "factories," all forms of shows, from vaudeville to major drama, were written, rehearsed, costumed and staged. A newly-established Soldier Actor Division created more than five hundred theatrical troupes, ranging in size from ten to over one hundred soldier-thespians. Involving over 12,800 men, these toured the AEF theater circuit, sharing the boards with the civilian talent from the United States and France. Thousands of additional doughboy-actors and performers were active in local areas, staging vaudeville shows, band concerts, plays, and musical revues, often written and scored by talented doughboys.[20]

Undoubtedly, the most popular programs were the musical revues, in which the men themselves were closely involved as active participants rather than as mere passive audiences. The problem of filling the many female roles required in these productions was variously solved. Female members of the several American contingents serving in France were recruited, and occasionally French girls performed. However, most of the roles were filled by doughboys, to the enormous amusement of their buddies. Preparations for their performances were elaborate, and with face makeup, wigs and gowns, some from famous modistes in Paris, and skillful coaching by professionals, wonders were accomplished.[21] Some of the shows featured professional actors then on duty with the AEF; others recruited only amateurs. They were all enthusiastically produced and received with much acclaim.

The main consideration was that some form of entertainment be available every night. Typical was the situation in the 352nd Infantry, of the 88th Division, where every evening, in each town billeting its troops, some amusement was provided, with a "literary night" being scheduled once a week. However, on the same night, on a rotating basis, a boxing match was held in an adjoining town, so as "to keep the regiment from becoming too cultural."[22]

Undoubtedly, the vast entertainment programs were most useful in helping to maintain morale among otherwise dispirited men eagerly awaiting shipment home. Those involved surely demonstrated that there was no incongruity between a soldier's wearing both a wig and a steel helmet, grease paint as well as a gas mask.

But the most impressive and far-reaching measures taken by the AEF high command to better the lot of its soldiers involved education. In this regard, the U.S. Army drew upon significant precedents and trends both within the service and in contemporary society. This era in American history was dominated by the ideas of Progressivism, the educational manifestation of which was known as progressive education, based on the views of John Dewey, the philosopher of pragmatism. The army reflected the times; professional training, which also grew out of hard military necessity, was well established by the time of World War I, and rapidly accelerated during its course.[23] In addition, Paragraph 449 of the Army Regulations stipulated that post schools, for the instruction of enlisted men, were to be an integral part of the military education system of the U.S. Army. Section No. 27 of the National Defense Act of June 3, 1916, elaborated, providing that in addition to military training, soldiers, while on active service, would be given the opportunity to study and receive instruction along educational lines "of such character as to increase their military efficiency and enable them to return to civil life better equipped for industrial, commercial, and general business occupations." These progressive education measures within the armed services reflected the views of the secretary of the navy, Josephus Daniels, and the secretary of war, Newton Diehl Baker. As one scholar has noted, under their leadership,

> The War and Navy departments became schools for social reform and personal improvement and sought to uplift America's soldiers and sailors, to indoctrinate them in such middle-class customs as frequent bathing, to regulate their drinking habits, to inculcate them with the principles of good government, and to make them literate.[24]

To men of such inclination, the problem of how to handle the large numbers of troops remaining in Europe following the end of hostilities was regarded as a substantial opportunity to further their mental improvement.

INTRODUCTION

As in the other programs outlined above, the YMCA was closely involved. In conjunction with its other welfare work, educational initiatives had been introduced. For example, both within the United States and overseas, the Y provided, among other subjects, for instruction in French and in world affairs and geography, in all of which the men developed keen interests. For the many illiterates and non-English speaking foreigners in the American army, English classes were set up, as was instruction in civics and elementary U.S. history. But the Y's leadership recognized that a systematic analysis of the AEF's educational requirements was needed. Consequently, Dr. Anson Phelps Stokes, secretary of Yale University and chairman of the American University Union, arrived in France to make a survey and draft recommendations for a viable, comprehensive educational scheme.[25] His plan, which was submitted to GHQ in February 1918, and promptly approved by Pershing, drew distinctions between the time of actual fighting and that following a cease-fire.[26] In the first period, Stokes recommended that only educational efforts that would directly contribute to the conduct of the war should be instituted. He suggested the continuation of instruction in the French language, in the history and the causes of the war, and related subjects, and the sending of lecturers into the field to supplement classroom work. A correspondence course program should also be instituted. All of these efforts were intended to help fit the American soldier for "the supreme military test of the coming year." It would help to "supplement the disciplinary value of military life," and to interpret for the doughboys the positive effects of "the quickening and broadening influence of living abroad." It was, in short, to help meet "the educational needs of a citizen army being fitted for effective warfare," and thereby contribute to making the soldiers "better fighters."[27] Even in this context, however, instruction should be limited to troops in training, at rest camps or in hospitals, with attendance placed on a voluntary basis. All of this instruction was to be supervised by the YMCA's Army Educational Commission then being set up in Paris. By the spring of 1918, the three-man commission was in place. It was headed by Dr. John Erskine, a professor of English at Columbia University. He also took charge of planning for work at the college and university level. Frank Ellsworth Spaulding, superintendent of schools in Cleveland, Ohio, was placed in charge of education below the college level, and President Kenyon Leech Butterfield of the Massachusetts Agricultural College at Amherst supervised agricultural, commercial, trade, and technical studies.

These men continued various educational programs to the degree that they could while hostilities continued, all the while planning for the time when soldiers could turn their attention from "triggernometry," to the more sedate study of trigonometry.[28] In the meantime, Stokes was aware that morale in the AEF would be improved if early publicity were given to the anticipated educational plans; the men would have this to look forward to after the end of hostilities.[29] What would then be needed would be instruction "so that our soldiers may return home more adequately trained for the business of life." This was intended "to fit . . . [soldiers] during the long period of demobilization abroad better to take their places on returning as workers in a modern democracy." This would include instruction in "the duties of citizenship," as well as in the ordinary branches of knowledge, vocational and industrial training. Thus, "if [the soldiers] can help make the world safe for democracy by a victory over German militarism and can then return home better fitted by education to take their part intelligently in solving our nation's great social and industrial problems, the United States will not have entered the war in vain." Therefore, Progressivism's ideals—which had created so much of the spirit of the times to begin with—would be further realized if the Army's educational plans could "make the citizen-soldier of larger value to the State."[30]

By October 1, 1918, the Y's Army Educational Commission had created numerous departments organized along specialty lines, such as agriculture, business, correspondence, and a department of citizenship instruction, among others. These were already providing instruction where possible, but were anticipating better times. Guidance as to how they would proceed came on October 31, 1918, when GHQ issued its first formal announcement of educational plans for the AEF in the form of General Orders No. 192.[31] Becoming effective on January 1, 1919, this order, based on the plans and procedures devised by the Y's Army Educational Commission, mandated a detailed educational organization within the AEF. It created the army post schools, at all establishments with at least five hundred soldiers, which, though provided for in Paragraph 449 in Army Regulations, had not been implemented during hostilities. Eventually, there would be about forty such schools in each division, making over a thousand throughout the AEF. A school officer was to be appointed in each regiment, division, corps and army, responsible for the organization of classes, the securing of classrooms and equipment, and for maintaining school discipline. Later, by General Orders No. 9, dated January 13, 1919, the army took over all instruction and administration of post schools, the Y's commission henceforth

INTRODUCTION

serving mainly in an advisory capacity.

Meanwhile, in December 1918, the army ordered to France Brigadier General Robert Irwin Rees, who, as chairman of the War Department's Committee on Education and Special Training, had organized the Students' Army Training Corps on American campuses. At Chaumont, he took over the army's educational organization, bringing considerable expertise, efficiency and much energy to the implementing of the impending AEF programs[32]

Further refinements of the AEF's educational programs came on February 13, 1919, with General Orders No. 30. This created divisional schools that were to teach courses at the high-school level, together with a list of fourteen designated trades. In the trade courses, men from both the post and division schools were often trained in connection with the work routinely carried out at the numerous large repair shops maintained by the army. The students were instructed by military personnel who were expert craftsmen and technicians. Some of the installations involved were the motor reconstruction park at Verneuil, the railway shops at Nevers, the motor construction park at Romorantin, the supply depot and Signal Corps shops at Gièvres, the ordnance shops at Mehun, the driver and mechanics school at Decize, and the remount station at Sougy. Over four thousand soldiers were eventually involved in this training.[33] Meanwhile, other educational programs were developed. These included the fielding of lecturers who singly, or in groups, traveled about delivering lectures to doughboys on such subjects as French history and art, and current political and social problems. Among these lecturers were special speakers sent throughout the AEF by the Citizenship Department of the Army Educational Corps. Both Pershing and prominent citizens at home were greatly interested in seeing that American soldiers were made into better citizens. The department emphasized such subjects as housing and community planning, public health and welfare, industrial problems, governmental organization and management, and foreign relations. The lecturers consisted of professors and other experts mainly recruited in the United States and sent to France specifically to lecture to the troops. The department also developed institutes, usually involving a team of four lecturers.[34] Especially popular and effective were other institutes devoted to agricultural as well as to business and commercial matters. Another innovation in education was the forming of clubs devoted to subjects of interest to soldiers who often organized and managed them with little outside supervision. Education was also rather more informally pursued through the Y's

Sight-Seeing Department, which set up tours to agricultural areas or factories so that interested soldiers might observe French farming and production methods. They also visited historical sites and art museums, and architectural field trips were arranged, to the château region in the Loire valley, for instance.[35] In addition, especially to meet the desires of soldiers stationed in isolated areas, the correspondence course system was steadily expanded.

The army also included a large number of men who desired to further their college or university educations. Recognizing their needs, General Orders No. 30 stipulated that detachments of selected soldiers, with high academic attainments, could be ordered to French and British universities for study. Cordial invitations came from schools in both nations, and after the armistice, the army implemented plans allowing about 7,500 to attend courses in France with an additional 2,000 or so in Great Britain. Fourteen French schools were involved, including the Sorbonne, where over 2,000 Americans studied. In the British Isles, forty-eight schools enrolled American soldier-students, including Oxford and Cambridge.[36] Their story will be told below in Chapter 8.

However, there were many more qualified doughboys than there were places in the British and French schools. To accommodate these, as well as those who did not meet the requirements for entry into the British and French schools, Chaumont decided to create an entire university for the AEF, beginning from scratch. In this, there was some guidance from the Canadians who also saw merit in providing educational opportunities for their men. One of their innovations was the establishment of several "Khaki Colleges," eleven of which were operational by May 1918. These were set up in England, the largest being at Witley. By May 1918, over 8,000 soldiers were studying in these schools, pursuing courses in general education, as well as commercial, agricultural, and engineering subjects. Each of these colleges consisted of a president, generally an officer of moderate rank in the army; a secretary, usually a member of the YMCA staff; and a faculty largely drawn from military personnel.[37] Guided by these precedents and the educational plans and initiatives already begun, the U.S. Army embarked upon one of its most remarkable undertakings: the founding of the AEF University at Beaune, in the Côte d'Or, in France's Burgundy region, renowned for its excellent wines and its rich tapestry of history.

INTRODUCTION

NOTES TO INTRODUCTION

1. *Stars and Stripes*, November 15, 1918; "The War Baby," vol. 1, no. 1, undated student newspaper at the Allerey Army Farm School, France. Even in the United States similar notions were entertained. One account recorded that a group of soldiers at Camp Funston in Kansas began to turn in their mess-kits and other equipment as soon as they received the news of the armistice. See William Howard Taft, *Service With Fighting Men. An Account of the Work of the American Young Men's Christian Associations in the World War*, 2 vols. (New York: Association Press, 1922), 1: 398.

2. There were also zones occupied by Belgium, centered on Aix-la-Chapelle; the British, at Cologne; and the French, headquartered at Mainz.

3. The quotes from Marshall are as cited in Donald Smythe, *Pershing: General of the Armies* (Bloomington: Indiana University Press, 1986), p. 249. There are useful accounts in *Stars and Stripes*, December 6, 1918 and February 28, 1919. See General Orders No. 207, GHQ, AEF, November 16, 1918, in United States Army, *United States Army in the World War, 1917-1919*, 17 vols. (Washington, D.C.: United States Army Center of Military History, reprint edition, 1988-1992), vol. 16, General Orders, GHQ, AEF, pp. 537-39, hereafter cited as: *United States Army in the World War*, vol. 16, pp. 537-39.

To be sure, some of the men lately arrived from the United States were only half trained and needed intensive drill and instruction. However, the veterans should have been spared. They had done their duty, and for them, the raison d'être of military life no longer existed. They had no desire to adopt a career in the army. They had endured a great deal in the national interest. This had now been served, and, as one writer observed, "the men's common sense rejected the futility of exercises and practice that had no significance except regarding a job now completed." Taft, *Service With Fighting Men*, 1: 170.

4. General Orders No. 241, GHQ, AEF, December 29, 1918, in *United States Army in the World War*, vol. 16, pp. 589-91.

5. C. Howard Hopkins, *History of the Y.M.C.A. in North America* (New York: The Association Press, 1951), p. 486. See also *Following the Flag: A Short History of the YMCA's Work with the Armed Forces* (Boston: Boston Armed Services YMCA, 1988).

6. The activities of the YMCA and the American Red Cross in the AEF were formally regulated by General Orders No. 26, GHQ, AEF, August 28, 1917. See in *United States Army in the World War*, vol. 16, pp. 60-61. This provided generally that the Red Cross was to engage in relief work, while the YMCA was charged with welfare programs involving amusement and recreation, designed to contribute to the social, educational, physical, and spiritual well-being of the troops.

7. See Pershing's Report to the secretary of war, Paris, September 1, 1919, in *United States Army in the World War*, vol. 12, *Reports of the Commander-in-Chief, Staff Sections and Services*, pp. 15-71. His discussion of leaves is on p. 68.

8. For details of the leave program, see Taft, *Service With Fighting Men*, 2: 142-62, and General Orders No. 14, GHQ, AEF, January 18, 1919, *United States Army in the World War*, vol. 16, pp. 613-17. Details of the leave trains are in Bulletin No. 14, GHQ, AEF, February 18, 1919, in *United States Army in the World War*, vol. 17, *Bulletins, GHQ, AEF*, pp. 194-95. Further discussions are in the troop newspaper *Stars and Stripes*, December 13 and 20, 1918, and January 17, February 14 and 28, 1919.

9. See also discussions in *Stars and Stripes*, March 14, April 25, May 2, May 9,

and May 23, 1919. Paris remained an unattainable goal for many men of the AEF, though not for all. Almost immediately after the armistice, Paris was filled with AWOL soldiers, confronting Chaumont with a major problem, at least partially solved by the stationing of fifteen hundred additional MP's in the French capital in December 1918.

10. *Stars and Stripes*, June 13, 1919. For further accounts of athletics in the AEF, see Taft, *Service With Fighting Men*, 1: 316-33, and 2: 26-54, and Wait C. Johnson and Elwood S. Brown, editors, *Official Athletic Almanac of the American Expeditionary Forces, 1919* (New York: Spalding's Athletic Library No. 77R, 1919).

11. *Stars and Stripes*, January 17, 1919.

12. Taft, *Service With Fighting Men*, 2: 38-41.

13. Ibid., 1: 324.

14. See accounts in *Stars and Stripes*, December 27, 1918; January 10 and February 21, 1919; Virginia Mayo, "That Damn Y": A Record of Overseas Service (Boston: Houghton Mifflin Company, 1920), pp. 245-52, 260-62; and Bulletins Nos. 6 and 28, GHQ, AEF, February 9 and April 5, 1919, *United States Army in the World War*, vol. 17, pp. 62-70, 245-46. There are numerous other bulletins reflecting the wide range of the athletics program in the AEF at this time in vol. 17. See also *Stars and Stripes*, April 11, April 25, May 16, and May 23, 1919.

15. Details of the Inter-Allied Games are in *Stars and Stripes*, January 17 and 24, April 11 and 18, 1919; Mayo, "That Damn Y", pp. 253-67; G. Wythe, J. M. Hanson, and C. V. Burger, eds., *The Inter-Allied Games, Paris, 22d June to 6th July, 1919* (Paris: n.p., 1919); and in a substantial documents collection pertaining to the Games in Record Group 120, Records of the American Expeditionary Forces (World War I), 1917-1923, subseries, "Records of the Chief Athletic Officer," Entries 345-349, National Archives, Washington, D.C.

16. AEF Entertainment Bulletin No. 1, January 28, 1919, is published as part of Bulletin No. 15, GHQ, AEF, February 21, 1919, in *United States Army in the World War*, vol. 17, pp. 195-200. See also *Stars and Stripes*, January 17, 1919.

17. Taft, *Service With Fighting Men*, 1: 619.

18. Ibid., 1: 622.

19. Ibid., 1: 623. These activities were coordinated by the Y's Entertainment Department in Paris, which after September 1918 was headed by the energetic A. M. Beatty, a theatrical manager before joining the Y.

20. Details of entertainment in the AEF are in Report of the assistant chief-of-staff, G-1, Brig. Gen. Avery D. Andrews, Chaumont, April 22, 1919, Chapter 15, "Entertainment," in *United States Army in the World War*, vol. 12, pp. 233-37, and Taft, *Service With Fighting Men*, 1: 619-36. Publicity for the entertainment programs was enhanced by the appearance of a regular column, "A.E.F. Amusements," in *Stars and Stripes*.

21. Some of these shows were described in *Stars and Stripes*, February 7 and 14, March 14 and 28, and May 16 and 23, 1919.

22. *Stars and Stripes*, February 7, 1919. Even an old-fashioned circus was formed at Bordeaux, its performers recruited from former circus people then in the army. It too made the rounds of the AEF show circuit. See account in *Stars and Stripes*, May 16, 1919.

23. Timothy K. Nenninger, *The Leavenworth Schools and the Old Army, 1881-1918* (Westport: Greenwood Press, 1978). Major sources for the study of the educational initiatives in general, and especially regarding the AEF University, are in Record Group 120, Records of the American Expeditionary Forces (World War I), 1917-23, subseries, "Beaune: American Expeditionary Forces University," Entries 408-420, National Archives, Washington, D.C. These records will hereafter be cited by

INTRODUCTION

folder, where applicable; box number; and specific entry number. There are other relevant documents in both the municipal archives and the municipal library, Beaune, France. See also the short study, edited by Lucien Perriaux, *Le Camp Américain de Beaune* (Beaune: Centre Beaunois D'Etudes Historiques, 1980).

24. Ronald Schaffer, *America in the Great War. The Rise of the War Welfare State* (New York: Oxford University Press, 1991), p. 100.

25. Taft, *Service With Fighting Men*, 2: 1-8. The American University Union represented abroad about 140 American universities, colleges, and technical schools. Its home office was at Yale University. Its European headquarters were at 8, rue de Richelieu in Paris, with branches in London and Rome. It had a twofold purpose: to aid American college men and their friends in war service, and to serve as a bond between the universities of the United States and those of the Allied nations. The union was often consulted by Y and army officials on matters pertaining to education in the AEF.

26. Anson Phelps Stokes, *Educational Plans for the American Army Abroad* (New York: Association Press, 1918).

27. Ibid., pp. 2, 3, and 6.

28. *Stars and Stripes*, May 24, 1918.

29. For example, *Stars and Stripes*, September 27, 1918, publicized the plans for education following the end of hostilities.

30. Stokes, *Educational Plans*, pp. 3, 6.

31. See General Orders No. 192, October 31, 1918, *United States Army in the World War*, vol. 16, pp. 508-10.

32. For Rees's activities in the United States, see Parke Rexford Kolbe, *The Colleges in War Time and After* (New York: D. Appleton and Company, 1919), pp. 60-81.

33. General Orders No. 30, GHQ, AEF, February 13, 1919, in *United States Army in the World War*, vol. 16, pp. 653-54. Additional accounts are in *Stars and Stripes*, March 28, 1919.

34. See undated report on the Citizenship Institutes in the AEF from their beginning until May 22, 1919, by H. R. Williams, who was in charge of organizing the lecturer teams, folder "Gen. files—Institutes," Box 1938, Entry 419.

35. Taft, *Service With Fighting Men*, 2: 23.

36. General Orders No. 30, GHQ, AEF, February 13, 1919, in *United States Army in the World War*, vol. 16, pp. 653-54. This order stipulated that all soldiers enrolling in British and French schools must agree to remain for the full term, unlike the provisions for men enrolled in the AEF University at Beaune. Further discussion is in Taft, *Service With Fighting Men*, 2: 16-17, and Stokes, *Educational Plans*, pp. 79-80.

37. There is a discussion in Stokes, *Educational Plans*, pp. 105 and 114; notes 5 and 6, p. 110.

CHAPTER 2:
Founding the University

ON FEBRUARY 5, 1919, Colonel Ira Louis Reeves, the commanding officer of the 137th Infantry, of the 35th Division, received orders at his headquarters at Sampigny, near St. Mihiel, to proceed to Chaumont, the General Headquarters [GHQ] of the AEF. Reporting to the army chief of staff, G-5 [Training], he conferred with Brigadier General Robert I. Rees, head of the Educational Sub-Section of G-5, about plans and developments for the launching of the AEF's own university.[1] Reeves was thereupon ordered to proceed to a former large AEF hospital center at Allerey, about ten miles southeast of Beaune, France, on February 7, to determine its suitability as a site for the proposed school. Rejecting it as inadequate for the larger university, though perhaps appropriate for an agricultural school, later in the day he inspected another hospital complex at Beaune. Satisfied that it would serve the purpose, on February 8, General Rees accepted Reeves's recommendations that it be selected, and immediately moved to create a campus there. On February 12, Reeves was detailed as the superintendent and commanding officer of the American E. F. University, and on the following day, General Orders No. 30, GHQ, AEF, confirmed the creation of the school. Formal authority to begin construction came on February 22.[2]

Colonel Reeves was chosen for the new position because of his experience as an educator, as well as an active duty officer with much experience in the army, in which he had served in the ranks as private, corporal, and sergeant, and all of the commissioned grades up to colonel. He was born at Jefferson City, Missouri, March 8, 1872. He studied at Purdue University and the University of Vermont, graduating from the latter in 1915 with a degree in civil engineering. His military career began in 1891 in the Missouri National Guard. He was commissioned a second lieutenant in the 17th U.S. Infantry on April 19, 1897. He was promoted to first lieutenant for his conspicuous gallantry at the battle of El Caney in the Santiago campaign in Cuba during the Spanish-American War. He later saw service during the Philippine Insurrection, being wounded six times. He was promoted to captain, retiring on November 11, 1902. As a retired officer, he became commandant of cadets and professor of military science at Purdue in December 1902. Much later, from October 1915 to August 1917, he became president of Norwich University, the military school of Vermont. This service was interrupted when the National Guard was sent to the Mexican border in July 1916, Reeves serving as the commanding officer of the 1st Vermont Infantry, where he remained until September. With the rank of major, on August 5, 1917, he once more returned to the active list of the United States Army, and was promoted to the rank of colonel in December of that year. During the early months of the U.S. involvement in the war, Reeves was on the staffs of the adjutant general and inspector general in Washington. Subsequently, his request for overseas service was granted, and he was assigned to the 64th Infantry, of the 7th Division. He was gassed in the St. Mihiel sector and sent to the hospital at Toul. Upon his recovery, he was sent to the Pont-à-Mousson area where he was attached to the 92nd Division engaged in receiving repatriated Allied prisoners. He was then ordered to the command of the 137th Infantry, 35th Division, on November 27, from which he was posted to the new university.[3]

Closely paralleling Reeves's activities in establishing the new school, on February 26, all division commanders in the AEF received a lengthy telegram with instructions as to how prospective students were to be selected and ordered to Beaune. The university would receive the first students on March 6. Officers and soldiers who were high-school graduates or with equivalent preparation were eligible to attend. Division commanders were in charge of the selection of students. Men whose organizations were ordered home would be released by the university in time to embark with their comrades if they desired. Those enrolled would receive full pay and allowances, and there was no charge for tuition, laboratory fees, textbooks, and supplies. Students would be awarded certificates of credit following successful completion of their courses. GHQ promised that a competent faculty would be assembled, that good accommodations and messing facilities would be provided, as well as good classrooms, laboratories, and library facilities. Indeed, "every effort is being made to provide an atmosphere typical of the American college."[4]

Reeves energetically turned to the myriad tasks of establishing the university. Numerous general orders, bulletins, and memoranda soon created the basis upon which the school would function. All of the land

2. General Robert I. Rees, Head of AEF Education.   111-SC-88400

controlled by the former hospital center was divided into two portions: the main campus and agricultural lands. The lands of the main campus were placed under the superintendent of buildings and grounds. Under his direction, the College of Agriculture planted and cared for the campus plants. The College of Fine and Applied Arts was in charge of landscaping. The College of Agriculture used some of the land surrounding the campus for demonstration purposes, planting plots with a variety of seeds.[5]

One pressing matter was obtaining a competent faculty. Some of those initially ordered to Beaune to serve as instructors were not of the caliber required for instruction in college courses, and Reeves repeatedly and urgently requested of GHQ that orders be sent to various AEF organizations seeking additional staff, with instructions that they use greater care in making selections.[6]

Reeves did not lack a philosophy to guide his administration. In the first place, he agreed with Pershing's desire that the university be open to all qualified officers and men of the AEF. The only limit would be the capacity of the institution to accommodate them. He was fully cognizant of the task of a "rather stupendous magnitude," that faced all concerned. Certainly, the many facilities and accessories that old academic institutions had acquired over the years would be impossible for the new school to obtain. All personnel concerned would therefore have to improvise and innovate, though every attempt would be made to secure everything required. As to the educational standards, these would "be of the very highest." Indeed, no effort was to be spared, he promised, "to provide those particular things which will assure for the A.E.F. University a high standing among the universities and colleges of America." Reeves realized that a careful selection of the students based on a review of the credentials of each and the arranging of qualifying examinations would be "next to impossible." It would therefore be necessary to accept each student's word as to his previous preparation. Indeed, he himself embraced the "rather modern idea," which was "possibly somewhat shocking to some of the older educational institutions," that entrance requirements to a college or university "should be solely that of the ability of the student enrolling to carry with entire satisfaction to his instructors the work assigned him." No doubt, some of the men would be unable to meet these requirements. Their identification and disposition would be left to the head of each college, as would the matter of general entrance requirements.[7]

Generally, Reeves understood that he and his staff were "here to serve," primarily the army and its soldiers, in accordance with the wishes of General Pershing that every effort be made to assist each soldier to return "to his home a better qualified man in every way to meet the obligations which will confront him in civil life, that he left when he enrolled as a soldier to fight his country's battles." These high ideals, and the plans to create a full-fledged university within thirty days—in fact, an institution that would probably soon be the largest university in the world—would have to be accomplished in the face of monumental difficulties. Reeves hoped that American colleges and universities would appreciate and sympathize with, as well as support, the school's efforts, principally by subsequently accepting a transfer of credits. Surely, he said, any "broad-gauge institution" at home could only applaud the army's efforts "for improving the minds and the morals of its fighting men," under the most adverse conditions.[8]

Though himself a liberal-minded academician, Reeves nonetheless never forgot that he presided over an *army* university. Several of his initial general orders clearly established this fact.[9] The code of conduct drafted for the students demanded that there be no smoking in classrooms; hats were not to be worn indoors; blouses were to be buttoned at all times, and no article not part of the prescribed uniform would be worn; and students were to obey promptly direct orders received from instructors. All members of the command were to be clean-shaven with hair properly cut, and with all apparel and equipment kept neat and clean. The appropriate military courtesies were to be rendered; it was especially noted that "there is but one salute and that is the one prescribed in Army Regulations." All organization commanders were to devote at least five minutes per day to instruction in the rendering of military courtesies, the wearing of proper uniforms, and soldierly bearing. In the barracks, bunks were to be properly and neatly made up, and "no loud, boisterous, profane or obscene language" would be tolerated. The movement of the students would be according to customary military practice as well. All units marching from one point to another, "on any military duty whatsoever," would proceed at quick time, in close order, and in columns of twos.[10] Another prohibition set the AEF's university apart from its civilian counterparts: all Greek letter fraternities were excluded. However, every encouragement was given to the establishment of literary, art, technical, and other societies, provided that they were open to all.[11] Thus, Reeves intended to establish a university with an attitude of broad-mindedness, liberalism, and inquiry, though within certain limits demanded by its military circumstances.

FOUNDING THE UNIVERSITY

General Orders No. 6, February 23, 1919 formalized the educational organization of the university.[12] The school was to be administered by military authority through the superintendent of the university, with the advice of the YMCA's Army Educational Commission and its experts. The superintendent issued all orders governing the nature and scheduling of courses, and the assigning of all administrative and instructional personnel. He was the chairman of the faculty and of the military staff, and an ex-officio member of the University Council and all other committees. The chairman of the Y's Army Educational Commission was designated as the educational director of the university, also serving as the chairman of the University Council. Directly under Colonel Reeves, this man was Dr. John Erskine. Born in New York City, October 4, 1879, he graduated from Columbia University with an A.B. in 1900, an A.M. in 1901, and a Ph.D. in English in 1903, and subsequently became noted as a poet. From 1903 to 1909, he taught English at Amherst; from 1909, he was professor of English literature at Columbia. Taking a leave of absence from Columbia, he came to France in January 1918, with Anson Phelps Stokes, who laid the foundations for the AEF's educational programs. For three months, Erskine worked at several huts of the French *Foyers du Soldat*, which were similar to those of the American YMCA. On April 1, 1918, Erskine took charge of educational work in the American Army under the general direction of Colonel M.A.W. Shockley of the General Staff of GHQ, the head of the sub-section in charge of education. When the YMCA's Army Educational Commission was formed, Erskine was appointed chairman. For a time, he returned to the United States to arrange for the selection and purchase of textbooks for use overseas, and helped arrange the shipping of massive numbers of books by the American Library Association to the AEF. Returning to France in November, Erskine and Brig. Gen. R. I. Rees, who had succeeded Colonel Shockley in charge of educational work at GHQ, completed arrangements with the French and British universities to receive American soldier-students as soon as these could be selected. Erskine subsequently remained in charge of the work in French and British universities. There is little doubt but that the "initiative, vision and energy" of Erskine contributed in large measure to the success of the university.[13]

The Y's Army Educational Commission advised the superintendent as to general educational policies and recommended to him the directors of the colleges and the heads of departments. The University Council consisted of the superintendent, the educational director, and

3. Dr. John Erskine, Columbia University.   111-SC-161035

other members of the Army Educational Commission as ex-officio members; the directors of the colleges; the registrar; and the head of the Saturday Course in Citizenship.[14] The Council recommended to the superintendent the selection of faculty and the nature and schedule of courses. The directors of each college in turn, recommended to the Council details of courses of the college and the selection of personnel. The directors, usually civilians, were the chairmen of their respective faculties and ex-officio members of all departments and committees. The assistant directors were always military officers.[15] General Orders No. 6 also dictated the organization of each department, when and how it was to meet, the committees each was to set up, and how each was to function.[16] Further details, including the order of business of the Council and each department, were set forth in General Order No. 14, of March 1, 1919.[17] The school's calendar, devised by the Council, specified that the first term of the university calendar would cover a period of three months from March 15, 1919, with recitations beginning on Monday morning, March 17. The term, based on the quarter system, was to conclude on May 31, a period of eleven weeks, exclusive of registration and examination week. All courses were to be in session from Monday through Friday, with recitation and lecture periods of fifty minutes each. Fifteen hours was the normal student load, with twenty hours permitted if a student was "clearly qualified to do so." Mid-term reports were required in all work indicating whether or not the student "has satisfied the instructor."[18] The grades recorded were to be: A=good, 90-100; B=fair, 75-89; C=poor, 60-74; and D=failure, 0-60. Subsequently, there was considerable preoccupation with grades and standards and with making the university generally respectable. Reeves assured Erskine that "no effort [would] be spared on the part of the military to provide those particular things which will assure for the A.E.F. University a high standing among the universities and colleges of America."[19]

Reeves was also the commanding officer, and the military side of the university was organized along traditional lines. Reeves's executive officer for a short time was Major Franklin Babcock, who was soon replaced by Major Livingston Watrous. Watrous supervised the nine student regiments, the auxiliary units, a company of German prisoners-of-war, and the major departments. These included the adjutant, the military director, the quartermaster, the superintendent of buildings and grounds, the surgeon, the military inspector, the judge advocate, the assistant provost marshal, the motor transportation officer, and the rail transportation officer. Watrous had, since early boyhood, planned a military career. After completing his elementary studies at St. Marks School in Southboro, Massachusetts, he entered St. Johns Military Academy at Manlius, New York. He graduated in 1908, but remained for two more years of postgraduate work and instructing. He was commissioned a second lieutenant in 1911, and assigned to the 27th Infantry. He was promoted to first lieutenant in 1916, captain in 1917, and major in June 1918. Commanding a battalion of the 807th Pioneer Infantry, he came to France in September of 1918. His organization was assigned to the First Army and went into action in the Meuse-Argonne offensive. When the university was first organized, Watrous was appointed military director, but soon became the university's executive officer.[20]

The university adjutant was Captain Waldo P. Hair. Under his command he had the personnel adjutant, the University Service Battalion, and the chaplains, as well as a staff of officers and enlisted men to assist him in his extensive administrative tasks.

The military director of the school—succeeding Major Waltrous—was Major Patrick J. Hurley. He was responsible for the students' welfare and discipline. He supervised their daily life directly through the commanding officers of the student regiments, of which there were nine, six at Beaune and three at the Farm School at Allerey. The regiments were composed of ten companies, with five companies comprising a battalion. The commanding officers of the regiments and battalions were tactical officers. These were assisted by student officers who were directly over the enlisted students. The military director also had under his office the athletic officer, Major William N. Howard; the entertainment officer, Captain M. C. Carroll; the supervisor of the regimental and other messes; and Troop "A," 15th Cavalry, which performed interior guard duty at the school. Hurley had been educated at Dartmouth University, receiving his B.A. in 1911. After graduation, for a time, Hurley was athletic director at the University of Puerto Rico. He entered the army as a second lieutenant in March 1913, and was assigned to the 3rd Infantry. He subsequently served in the Canal Zone with the 10th Infantry. He came to France in April 1918, as a member of the 7th Infantry, 3rd Division, and saw action at Belleau Wood and the Battle of the Marne. He also served as an instructor in the Army Candidates' School at Langres. Late in February 1919, he was detailed to the university in command of the 8th Provisional Student Regiment, becoming military director soon afterwards.[21]

The university military inspector was Colonel S. Field Dallam. He

FOUNDING THE UNIVERSITY

exercised general supervision of the military phase of the university's life. He was also responsible for much of the liaison work with the French, such as guiding French visitors about the campus and coordinating various French-American activities involving the university. Dallam was born in Philadelphia in 1874 and was graduated from West Point in 1896. Commissioned as second lieutenant, he was assigned to the 1st Cavalry stationed at Fort Riley, Kansas. During the Spanish-American war, Dallam saw action in Cuba with the 8th Cavalry. In 1901, he was promoted to first lieutenant and sent to the Philippines with the 5th Cavalry. He remained in the islands for three years, and returned in 1909, serving two more years there with the 12th Cavalry. He subsequently served two years on the border with Mexico with the 7th and 8th Cavalry. In June 1918, he was assigned to the Inspector General's Department. Later, arriving in France, he served with the 1st and 5th Army Corps. He was assigned to the university in April 1919. For a brief period, when Colonel Reeves was absent in early May, touring the Third Army's area in Germany, where he inspected the post and divisional schools, Dallam was in command at Beaune.[22]

The university's judge advocate was Captain J. D. Matthews.

The Motor Transport Corps consisted of M.T. Company No. 477 and several men on detached service from the Service Battalion. Captain L. A. Bonner commanded the unit. It had been stationed at Base Hospital No. 12, and when it was closed, the MTC personnel were scheduled to return to the United States. The conversion of the camp into the AEF University resulted in a postponement of their return home yet, "in spite of their disappointment, they stuck to the job with a renewed vigor and cooperated in every way to make the university a success," an attitude Reeves warmly praised.[23]

The quartermasters at Beaune were commanded by Major Andrew J. Bush. The disbursing quartermaster, was Major F. E. Barnum. The school's sanitation officer was Col. Joseph Herbert Ford, an expert in the field of army sanitation administration, who had authored numerous textbooks on the subject.[24]

The University Service Battalion numbered twelve hundred men, and was under the command of Captain Hisketh, of the 164th Infantry. It was charged with the general upkeep and operations at the camp. The battalion included eighty mess sergeants and numerous skilled cooks. Many of the battalion's mechanics and drivers were temporarily attached to the Motor Transport Corps for duty at the school camp.[25]

Captain H. B. Toman, assistant provost marshal, was in charge of the company of military police numbering 175 men, 60 of whom were stationed in Dijon, the nearby town where many off-duty students visited.[26] They also regulated the traffic to and from the university and maintained order in Beaune. Originally of the First Army, they had "directed traffic in the Argonne." They experienced little trouble from men at the university, as the MPs found "that the type of the men are above the average in France." The interior guard and other police duties at the school were accomplished by Troop "A," 15th Cavalry, commanded by Captain E. P. Gofnell. He was also in charge of the detachment of German prisoners-of-war, on "detached service" with the troop, though their labor was directed by the superintendent of buildings and grounds.[27]

An engineer detachment was part of the school's military establishment. Commanded by Major Richard Brooke, the engineers were responsible for building the roads throughout the camp. They also constructed and renovated several hundred concrete and wooden barracks, built the theater and other prominent buildings, and installed a drainage system.[28]

Care was taken that the military phase would not interfere with educational work, which was regarded as the prime mission of the base, even though the students were marched to their colleges and placed in charge of their educators by their regimental officers at 8:00 A.M. and at 1:00 P.M. five days per week. In the evenings, the hours from 8:00 to 9:45 were regarded as study hours; those not desiring to study were required to observe quiet. At all of the morning and afternoon recitation periods, and during the evening study hours, "education rules supreme," and no military duties were required. "The only thing military [the student] has to do is to remember that he is still a soldier, whether in classroom or barracks." At all other times, beyond the educational periods, however, the soldier-collegians were under the direct control of the military branch of the institution. In this way, both military and educational goals were pursued concurrently with obvious great success, bearing out that the ideas of Reeves and others, such as officials on the Army Educational Commission, who had fathered the university, were sound. Indeed, many experienced military men were of the opinion that the university as planned could not work, especially when privates were to instruct majors and company officers sat in the classroom with the men under their command. But General Pershing, when he twice inspected the school, praised the men for their soldierly appearance, military bearing, and efficiency. Thus, even though the program used at the university was absolutely new, and "no army [had] ever tried out such a plan of education before," it is clear that the system was immediately successful and it remained necessary only to fine-tune the enterprise.[29]

The library was among the school's impressive departments. The chief librarian was Luther L. Dickerson, of the American Library Association's Headquarters Staff in Washington, and formerly of Grinnell College. The library eventually consisted of almost thirty thousand volumes provided mainly by the American Library Association's Library War-Service. The basis of the collection was the "ALA Educational List." Books were classified by the standard Dewey Decimal Classification, and were similar to those found in the average American college or university, though they were the latest editions. Special purchases from French and English book markets placed in the Beaune library many books not yet found at American schools. The library was housed in three large, connected buildings. Several reference rooms were needed to handle the approximately five thousand books reserved for required or supplementary reading. Accommodations were provided for seating about fifteen hundred readers at one time. The staff consisted of twenty-four librarians, several of whom were trained and experienced personnel from the ALA. Almost one hundred students worked as assistants as well. By mid-May, the per diem circulation was about fifteen hundred volumes per day.[30] There were no charges or fines assessed, and books were lent to all members of the command, regardless of whether they were enrolled in classes or otherwise assigned.[31] The library was officially opened on March 12, but the initial press of the student-readers was so great that the library had to close almost immediately, until it could be expanded to handle the crush.[32] Dickerson, having successfully organized the university library, was then sent to the headquarters of the U.S. Army of Occupation in Germany at Coblenz. His replacement was F.L.D. Goodrich of the University of Michigan Library.[33]

The university had four registrars who were put in charge, not only of the university, but of all the school systems of the AEF. The university consisted of the following colleges: Agriculture, Fine and Applied Arts, Business, the Cadet College, and the Colleges of Correspondence, Education, Engineering, Journalism, Law, Letters, Music, and Science and the Departments of Medical Sciences and Citizenship.[34]

The Cadet College was one of the school's special creations. It was for the exclusive use of a detachment of West Point candidates.

SOLDIER-SCHOLARS

Following its own program and routine, it began its sessions on February 15, and lasted for four weeks. Its commandant was Captain Isham Henderson, of the Field Artillery, with Roy Everet Warren as its director. A civilian, Warren had been an instructor in education in the Extension Division of the University of California at Berkeley, and the principal of the Whittier School in that city. The college provided instruction in those subjects which were required of applicants taking the entrance examinations for West Point, and followed the curriculum laid down by the Academy. Two hundred and ninety-eight students were registered initially; the preliminary tests reduced the number who took the formal examination on March 18-20 to 188. Subsequently, while awaiting the results of their exams, which had been sent to Washington for scoring, the cadets were transferred to the regular courses of the university, where their studies conformed to the first-year curriculum at the Academy. In May, word came that only nineteen had passed. The successful candidates were ordered to report to West Point by June 13, 1919. All others returned to the United States for their discharges, or to their respective organizations. Those sent to the United States left Beaune on May 20, bound for St. Aignan and from there to a base port for embarkation.[35]

Another specialized entity was the Correspondence College. It was in essence the extension department of the American E.F. University. It provided educational opportunities for all members of the AEF who for any reason could not attend a suitable post or division school, or who were not qualified to attend any of the English or French universities or the University at Beaune. It expanded the correspondence course system already in operation. Earlier, students applied to the Y's Army Educational Commission at 76, rue du Faubourg St. Honoré, Paris, for the courses, which had steadily grown in number and popularity.[36] With the creation of the College of Correspondence at Beaune, new course applications were henceforth sent there.[37] J. Foster Hill, a Harvard graduate from Scranton, Pennsylvania, a town noted for correspondence schools, was the educational director of the college. Engaged with the International Correspondence Schools from 1908 to 1915, he was thoroughly experienced.[38] Enrollment figures for correspondence courses climbed rapidly, and by the end of May, had reached almost seven thousand.[39] Fifty-seven subjects were taught by mail, with salesmanship and personal development, civics and citizenship, farm management, the gasoline automobile, shorthand, business law, and foreign trade being among the most popular subjects, though the men could also take courses in strength of materials, applied electricity for practical men, and poultry husbandry, among others, though one soldier was told that his request for a course for railroad conductors could not be met because none was offered. The courses reached students in the navy on the high seas; soldiers stationed in the Army of Occupation in Germany, where over two thousand were enrolled; in England; and in "unheard [of] nooks" in France. There was a staff of fourteen instructors and a mailing crew of seven, aided by large details of students from the other colleges of the university. Though most students requested two courses at a time, a scarcity of materials and books limited each to one course. Many students had to return their materials and books unused because they were being shipped home. A considerable amount of correspondence was taken up with these matters. Often when the college could not supply a desired course, the ALA was instructed to send on to the soldier some appropriate book on the subject that interested him. The ALA supplied many of the texts and technical books required, though the Y also supplied texts and materials. In addition, many bound correspondence course pamphlets and syllabi were obtained from the Commonwealth of Massachusetts Department of University Extension of the State Board of Education. Some of these courses were: elementary applied arithmetic, electric wiring, practical calculus, concrete and its uses, and advertising, among many others. The Correspondence College received about ninety-five thousand pamphlets and syllabi devoted to forty-eight courses from Massachusetts. However, many of these were based on textbooks not available in France, and others duplicated courses that were already being offered. When the college ended its activities, much of the Massachusetts material was distributed to the students for their individual use.[40]

There is no doubt that the advertisements and articles about the correspondence courses in the *Stars and Stripes* and in other troop newspapers contributed to the high enrollments. Some ads included application blanks for the convenience of their readers. Many of the soldiers applying for courses, often on stationery with YMCA, Salvation Army, or Red Cross letterheads, mentioned that they had noticed the ads and articles. They readily responded to such ploys as: "While waiting for your good ship homeward bound why not do some school work by the correspondence method?"[41] There were no charges to the soldier. What did he have to lose?[42]

Encouraged by the widespread advertising, applications poured in. Many of these reveal the strong desire and gratitude of the men for opportunities for study. Private Robert L. Branch, barely literate, wanted to take a correspondence course in arithmetic, world geography, or the Gregg System of Shorthand. However, he noted that his first "disire" was to study the "Holly scripture," which was "my calling," but unfortunately, no such course was available.[43] Private M.A. Church, who had worked on freight boats, wanted courses in arithmetic and steam engines. He concluded: "I have practically no knowledge of arithmetic, Small knowledge of engines, A very large amount of energy to learn."[44] Private James C. Dickson, as was true of so many others, wanted a course in agriculture. "I like agriculture better than anything else that I know of. . . . Am a lover of stock and have my home in Middle Georgia," he declared.[45] Private H. Fenski, of the 11th Infantry, also wanted a course in farm management. His former occupation was that of baker, but "I do not like indoor work any more, especially since I have improved my health by being out-doors a whole lot. It is my earnest wish to go into scientific farming when discharged from service."[46] Private Charles P. Gassman, from "Somewhere in the Army of Occupation in Germany," wrote on March 5 that only "God knows when we will get home." While in Germany, he wanted to study English, grammar, and arithmetic. "So Id like to know wheather or not I could secure an education through correspondence school. I surely do hope that I will not go home and be as ingnorant [sic] as a Russian Soldiers," he concluded.[47]

Hill was gratified that the written lessons received from the students had been of "exceptionally high grade," and he regretted that the program had to be cut short. Had the school continued for a few months longer under more stable conditions, he was certain that an even more enviable record would have been made "that would surely prove the worth of the correspondence method of instruction," which seems to have been successful, in any case.[48]

Another major innovation at the university was the creation of the Course in Citizenship, presented by the Department of Citizenship, which ranked with the other colleges of the university. The director of the program was Lt. Col. William Freeman Snow, who was also head of the College of Science at Beaune. In civilian life, he was a professor of hygiene and public health at Leland Stanford Junior University, and a member of the Health Commission of California. During the first part of America's involvement in the war, he was an executive officer on the staff of the surgeon-general in Washington. Sent overseas, he was assigned to the staff of the chief surgeon in Chaumont as a medical and surgical consultant, prior to his transfer to the university. While every college and every department at Beaune was attempting to equip its students with a training that would enable them to make the transition from being a "part of General Pershing's gigantic fighting machine to being valuable citizens," none was more closely engaged in "making an effort to present to every man the outstanding questions which every citizen must answer if his community is to progress." While solutions to outstanding questions, "which every citizen must answer if his community is to progress," were not given, emphasis was placed on how the citizen should study them and determine correct solutions for himself, "which he is prepared to support." The method for accomplishing these ends was "focusing special attention on a selected group of problems in the all-encompassing realm of CITIZENSHIP."[49]

It was originally planned to organize the students into sections of two hundred each for lectures, and then have an hour's discussion. Several sections were then to be combined for an hour of motion pictures and lantern-slide presentations. But the lack of a sufficient number of large classrooms prevented the carrying out of this plan. Consequently, it was necessary to give lectures in sections numbering several hundred. These presentations were made on Saturday mornings—later shifted to Tuesday evenings. All students were required to attend the program, which consisted of an address, followed by an hour's discussion, with a third hour being devoted to lantern slides or a movie. The address, as prepared by the week's designated speaker, was delivered prior to the presentation to an assembly of instructors drawn from the general faculty, who in turn delivered the paper to the several groups of students. The discussion hour followed, under the direction of student leaders, with students then being formed into larger assemblies of about twenty-five hundred or so for the visual presentations. All those leaving the lecture halls were given copies of the printed lectures, though limited printing facilities initially hampered this. However, as soon as the print shop increased its capacity to produce the vast quantities of printed material that the university required, many of the lectures were printed and distributed to the students. Copies of these lectures were also sent to state and public libraries and to colleges and universities in the United States, as well as to French colleges and universities. Students were also provided with pamphlets, special papers, and other materials. To the twelve-week, three-hour sessions were added twenty-four hours of assigned reading and the preparation of a paper on some selected and approved subject of citizenship. A student might elect not to meet these requirements, but no credit was given for attending the three-hour weekly sessions, though attendance was compulsory.[50] In this course, one of the three major goals set by Erskine at the beginning—that of the development of citizenship and civic responsibility—was addressed. Indeed, Snow desired to demonstrate the course's success at Beaune in the hopes of inducing the colleges and universities in America to adopt similar ones. He did not believe that the academic institutions at home were giving "adequate attention to preparing their students to take an intelligent and helpful leadership in solving the problems of health, general welfare, and social and industrial progress which arise in every community in which they may become resident." Snow was subsequently pleased with the course's success at Beaune, and was confident that its value would be "increasingly evident as the men return home and take their part as leaders in citizenship affairs in civil life."

A second part of the course consisted of a series of daily lectures for which the students received five hours of credit on an array of subjects in the fields of science, literature, sociology, economics, and politics. These were presented by lecturers from various colleges of the university.

A third part of the course consisted of a series of popular lectures in art, science, literature, engineering, music, and current history given from time to time in the evening for those desiring to attend. Attendance at these presentations was on a voluntary basis.

When the course began, fifteen major popular lectures were planned, but the press of time dictated that only eleven were delivered. The first, "Society as a University," was presented by John Erskine. He outlined the purposes of academic institutions with particular reference to the school at Beaune. Another, "Educate America," by Frank Ellsworth Spaulding, developed a detailed program for the advancement of public education in the United States. In his view, all citizens should be provided with essential elementary knowledge, training, and discipline, which would contribute to occupational efficiency and inculcate a sense of civil responsibility in all citizens.[51]

One lecture in the series was by Grosvenor Atterbury, an architect, on city planning and community housing.[52] Lt. Col. John Price Jackson, head of the Industrial Department of the Citizenship Department of the Army Educational Corps in Paris, lectured on "Industrial Problems." Presented with a distinct bias, Jackson's lecture stressed that idleness was "possibly the most vicious disease which the human being can contract." This being the case, workers should realize that "labor under proper conditions was a blessing and idleness a curse." He deplored labor unrest and agitation "based upon wanton destruction or anarchy," which was calamitous. He recognized that both large corporations and labor unions alike were great concentrations of power that had to be guided carefully to prevent injury to the public interest. He insisted that differences between capital and labor could be settled in an orderly manner without either strikes or lock-outs. He hoped, finally, that the continuing "terrible toll" of industrial accidents and sickness might be addressed and eased.[53]

Also in the series, Colonel Joseph Herbert Ford, the chief sanitation officer at the school, lectured on public health, while William J. Newlin, in civilian life a professor of philosophy at Amherst College, focused his paper on the "Principles of Democratic Government." Extolling the virtues of democracy, he noted that kings and emperors had in their day of rule "rendered great service" to the world, but "their day is over." Because few men can rule well as supreme leaders—Lincoln was an exception, he declared—"the common sense of justice and liberty and equality is the only safe rule of conduct." Happily, the democratic era had now arrived.[54] Herbert Hoover lectured on "The World's Food Supply," while other presenters discussed, among other subjects, the elements of citizenship, public safety and welfare, foreign relations, and art and the citizen. Snow himself closed out the course with a special address, "The Principles of Citizenship Applied," in which he sought to set the capstone on the citizenship program.[55]

The GHQ, AEF, in its General Orders No. 9, January 13, 1919, had stipulated that the men at every place with a constant population of five hundred or more soldiers were to have access to a post school. The attendance was on a voluntary basis, except for illiterates and non-English-speaking soldiers who were to be directly ordered to schools by commanding officers. The subjects to be taught included common school courses, modern languages, U.S. history, histories of modern nations, civics, and citizenship at the elementary school level. Once entered, the courses had to be completed. Attendance was required. If the men were transferred or duties interfered, a record was kept so that when the opportunity presented itself again, the student would be able to resume his studies.[56]

There were also the division schools, which concentrated on vocational courses and academic and commercial subjects of high school levels. These included tailoring, barbering, baking, carpentry, cobbling, telegraphy, wireless telegraphy, telephone construction, laundry operation, horseshoeing, land surveying, road construction, mechanical drawing, trigonometry, algebra, salesmanship, economics, and advanced language courses in French, Spanish, Italian, and German. The divisional schools in Luxembourg, for example, stressed agriculture, with an emphasis on European intensive agricultural methods. The division schools soon began to appear in many sites in France, Luxembourg, and Germany. Many buildings were taken over and revamped as needed, some being abandoned factory buildings, warehouses, school-houses, halls, and casernes.[57]

The American E.F. University also created post and division schools, with Major L. Frazer Banks, of the field artillery, serving as director. Intended as model schools, they were charged with supervising the activities of all post and divisional schools throughout the AEF. This part of the work was supervised by Dr. Frank E. Spaulding, of the Army Education Commission. In each army area, and in each SOS Section, there was a general superintendent. All together, about 350 field personnel, and about 150 lecturers, institute organizers and instructors, and from 6,000 to 7,000 teachers, mostly drawn from the army, managed the program. About 10 percent of the men in the AEF, ultimately more than 200,000, were enrolled in regular classes in the post and divisional schools.[58]

The Beaune Post School also provided providing educational opportunities for the men on duty with the various organizations connected with the university. Every man on the post filled out a questionnaire stating his previous education, vocational experience, plans, and subjects in which he was most interested. The men who expressed a desire to take school work were divided into three groups. Those who were high school graduates could get off from work for some free time, enabling them to take a couple of the regular courses offered in the university. Similarly, those who had finished elementary school could take courses offered in the Division School. For men of the third group, those who had not finished elementary school, special classes were started in English, arithmetic, and citizenship. The men were divided by race, and 2nd Lieutenant Felix A. Scott was appointed associate principal and put in charge of all the Negro classes. Second Lieutenant John M. French was appointed associate principal in charge of all the white classes. Those men who were suspected of being unable to read and write English were sent by their commanding officers to be tested by a group composed of the instructors in the College of Education, and in the post and division schools. Their ability to read was assessed and they were assigned to appropriate classes. They were required to attend school from 1:20 to 4:20 P.M. five days a week. In addition to these courses for illiterates, classes were also begun in arithmetic and in English for more advanced elementary school pupils. The problem of instructors was a difficult one. However, a number of university students were found who were experienced teachers. These were willing to teach for an hour each day in the Post School. They were supervised and assisted by experienced instructors in the Army Educational Corps. In the classes for blacks, black students in the university were used as teachers also, though as these men left to return home with their organizations, they were replaced by white teachers. Something over two hundred native-born and foreign-born white students were accommodated in the University Post School classes, and about one hundred black students were enrolled. Banks noted that the results of the work had been good, and the men were proud of their progress and demonstrated a great interest in making further advancement.[59]

The Division School, whose principal was A. C. Davis of the Army Educational Corps, was intended for students who were high school and elementary school graduates, who were not otherwise qualified for regular enrollment in the university, and who could attend classes either on a full- or part-time basis. In addition, the College of Engineering desired that some of its practical vocational subjects be taught in the Division School as well, so that it could concentrate on more theoretical work. The school, which began its classes on April 7, accordingly offered vocational, commercial, and academic subjects commonly found "in a first class high school" in the United States. Among the academic courses taught were English, Latin, French, algebra, geometry, physics, chemistry, economics, and American and European history. Some of the commercial courses were arithmetic, bookkeeping, typewriting, commercial law, and shorthand—both the Pitman and Gregg methods. The vocational courses included auto repair, gasoline engine operation, mechanical drawing, telegraphy, telephone installation and repair, carpentry, surveying, and airplane repair. By the end of the school, 316 students were registered full time in the Division School, with 755 taking one or two courses as part-time students. The Division School and the Post School were also used by the College of Education for its observation classes.[60]

Reeves also ordered that every facility be offered to enlisted men who were on duty at the university to take night courses. To provide stability, those who volunteered to take such courses were required to complete at least one unit of twelve classes. Reeves promised that "the best instructors available will have the work in charge."[61] The night school was placed under the supervision of William Henry Lough, director of the Business Education Department. It was held "for the benefit of all personnel connected with the university, who are not enrolled as regular day students." So far as practical, students so enrolled would be excused from any military duties that might interfere with the regularity of their attendance.[62]

The daily routine at Beaune was closely scheduled regarding both the military activities and academics.[63] The weekday began at 6:00 A.M. with first call, followed by reveille at 6:15. Breakfast was at 6:30, followed by drill assembly at 7:20. The period from this time until 8:00 was spent in intensive instruction in close-order drill or military athletics. Further drilling was done each afternoon as regiments paraded according to a rotation schedule, one each weekday. Office routines began at 8:00 A.M., continuing to noon, resuming at 1:00 P.M., until 4:30, unless additional time was required for "the prompt and efficient transaction of official business." The classroom schedule began at 8:20, with traditional fifty-minute classes being scheduled throughout the day until 4:10 P.M., when classes ended. First call for parade for the regiment designated for that day was 4:30, with supper at 5:45. A study period was mandated at 8:00 to 9:45, with tattoo at 9:45 and taps, with compulsory roll call, at 10:00 P.M. Saturday morning classes were scheduled, for a time, notably those of the Citizenship Program, ending at 12:10 P.M., when the weekend routine commenced.

Before the university could begin operations, however, many buildings had to be renovated or built. Eventually about 200 buildings were completed or altered, with 175 additional structures erected.[64] To accomplish these ends, three regiments of black stevedores and a battalion of engineers were detailed to the work. In addition, about 400 prisoners-of-war were employed in construction and maintenance chores, living in a dormitory stockade that they had built for themselves.[65] But the construction troops assigned to the tasks were not sufficient in numbers to accomplish what was needed in time for the beginning of classes. Accordingly, students and faculty members were ordered to assist in getting the facilities ready. No doubt this came as a rude surprise to many. Nonetheless, the men turned to their assigned tasks. A detailed plan was worked out for this purpose.[66]

When, despite these efforts, a crisis was reached regarding getting the university ready to open on schedule, extraordinary measures were taken. Among other things, sixty railway cars loaded with supplies and materiel for the school had arrived, with ninety more expected, and numerous small but important projects remained incomplete. To ease matters, March 20 was set aside as a general construction and work day, and every officer, member of the faculty, and student was utilized to accomplish as much as practical during that time. It was "believed that if everyone will perform his part, that in a period of eight hours the crisis will be passed." The effort was successful, and the school proceeded to its formal opening on schedule.[67]

The university at Beaune aspired to be the center of all the educational efforts then underway within the AEF, with a zeal to make available widespread and intensive educational opportunities to all qualified military personnel desiring them. The school could not easily administer entrance examinations, and indeed adopted a philosophy, which Reeves and Erskine shared, that the only criterion for enrollment should be ability to accomplish the course requirements. This meant that a rather free-and-easy enrollment ensued. Yet some of the prospective students were plainly unqualified for university-level work. Reeves decided that such students should not be abandoned, however. He enabled them to remain at the university, and take prep or vocational courses in the Division School.[68]

The question arose whether civilians or women might enroll as students at the new school. Reeves decided that as "the American E.F. University was established for the officers and soldiers of the American Expeditionary Forces," it was not advisable to permit civilians to enroll. Women of the Army Nurses Corps, the American Red Cross, and the YMCA, stationed at Beaune, were permitted to attend classes, though they would not be formally registered as students. Furthermore, Reeves added, "attendance at classes must in no way interfere with their duties."[69] Once the university got underway, refinements on many of the earlier orders and instructions were made, and new ones appeared as problems arose. Classroom order and decorum required attention. One general order forbade lounging in classrooms, "such as rearing back in chairs or placing feet on seats." The proper wearing of uniforms was required; the wearing of hats in the classrooms and smoking had to be repeatedly forbidden. Students were required to police up after each class and leave the room in an orderly manner—if necessary, by formally marching in and out.[70] Lapses in military courtesy, especially regarding proper salutes, were a matter of perennial concern, and because cautions to the soldiers regarding this matter did not appear to be effective, the military police were ordered to arrest and return to camp any soldiers neglecting to render the proper salutes while on leave or on a pass.[71] Other orders concerned leaves and passes. The tendency was to make these easier to obtain as time went on, a sign of generally good discipline on base.[72]

If leaves and passes became progressively easier to obtain, certain procedures mandated by GHQ, AEF, in Chaumont, never varied. From the time that Americans arrived in Europe until their return to the United States, General Pershing and other concerned commanders waged an unceasing battle against venereal disease.[73] Beaune was no exception. All men returning to camp were questioned as to whether they had been exposed to venereal disease. If so, they were directed to the nearest venereal prophylactic station. Men who returned to camp intoxicated were ordered seized and taken to the prophylactic station, where the person in charge administered remedies. Reeves further demanded that these men be confined to the guardhouse. An unending flow of information about venereal diseases and the value of prophylaxis, penalties, and locations of the stations on the base was disseminated, as well as exhortations for the men to keep "clean."[74] The hospital staff at the university consisted of 25 officers and 191 men with 90 nurses. These staffed two camp hospitals, with two laboratories, a sanitary department, and a special VD Prophylaxis Department. The latter maintained nine VD stations at the Beaune campus with three at Allerey. Their duties were vigorously pursued. Men admitted to hospital with VD were followed up and tried at courts martial. Each man treated at pro stations was interviewed in an effort to ascertain the name and address of the prostitute, who was no doubt involved, "with the idea of either excluding her from the town or having her treated should the soldier develop Venereal Disease." Any infected prostitute was examined by French authorities or university hospital authorities and, if infected, was treated under French jurisdiction. Organizations having a high venereal rate were not given pass privileges until their rate was lowered to fifty cases per thousand, or less. Leaves were not given to men known to have VD, and many were court martialed. The usual punishment was reduction to private if the offender were a corporal or a sergeant, and a fine on the order of $10.00 per month for one month, or $7.50 for two months.[75] Printed circulars about the dangers of VD and their consequences, both legal and medical, were issued to the commanding officer of every organization. Lectures under the authority of the chief surgeon's office, GHQ, AEF, were given to every person in military service at the university. The VD rate at the school was about the same as the average rate within the AEF at large.[76]

FOUNDING THE UNIVERSITY

Not only were generous leave and pass provisions enacted, but many of the officers on duty at the university, except those in command of troops, could obtain permission to live in Beaune. A few enlisted men were also able to take advantage of this privilege. There were restrictions: permission to reside in town in no way relieved them from any of their duties; they had to have formal written permission to reside in town from 1st Lieutenant Charles C. Benson, who was in charge of these arrangements. Benson had to inspect and approve the prospective quarters. To combat excessive charges by French proprietors, officers were ordered not to pay in excess of fifteen francs per day for room and board, or one hundred francs per month for billets alone. A French committee, "The Beaune Committee of French Homes," supervised by the town's sous-préfet, assisted in locating acceptable quarters. On the average, just over 150 university personnel resided in Beaune, most of whom were officers.[77] To be sure, a rather large number of student-officers, among them the five assigned to each company of the provisional companies, were considered as in command of troops and consequently unable to reside in the city.[78]

Reeves took advantage of the students' unusually favorable situation to assist in maintaining discipline. One bulletin containing basic guidelines for conduct and attitude was accordingly widely circulated. It reminded the men that they were a picked corps made up of personnel with past scholastic achievements, and were being accorded an honor eagerly sought. Consequently, each member, "aware of his individual responsibility for the good name of the Army and the Nation," would recognize how vital it was to maintain "an exceptional [sic] high standard of personal conduct." The new school should be regarded "as a real university," with as few of the restrictions of army post life as possible. To these ends, and "pending evidence to the contrary," it would be assumed that all the enlisted men merited good-conduct passes and could therefore travel to and from Beaune without a formal pass, as had hitherto been accorded to officers only. No student was permitted in the city without written authority on any day during hours when the school was in academic session. The students could go into town if study periods were not in force, especially Friday and Saturday nights up to midnight. Officers, too, were to conform to the highest standards of conduct, being held "strictly accountable for any breach of the spirit as well as of the letter of these regulations."[79]

Another bulletin, entitled "Obligations of Students To The University," indicated how the students might further repay the army for the privileges they enjoyed. One practical way would be for each to devote part of his time to the university's physical needs. Administrative troops were not available to do everything; the students were to fill in by giving an hour a day for the maintenance of facilities, and the care of classrooms, laboratories, grounds, and athletic fields. In addition, beautification projects would be carried out according to plans drawn up by the College of Fine and Applied Arts. Each college was assigned an area on the post as its responsibility. Detailed plans were made, setting forth the precise duties of each student enrolled. An inspector had been appointed to oversee the work. Reeves hoped that prizes would be offered later in the form of certificates to those colleges whose areas were considered the best kept and the most artistic, the judges being selected from personnel not connected with the university. Reeves intended that by using student help, the school might be made "a model of cleanliness and artistic arrangement."[80]

In other ways, there were appeals to the honor of the students. It was generally understood that the hours between 8:20 A.M. and 4:10 P.M., less the noon hour, were to be devoted "entirely to classroom work or preparation for classroom work." During any vacant periods in a student's schedule, students were to utilize the time for study, seeking out laboratories or vacant classrooms, except for those of Provisional Regiment No. 11, who might return to their quarters to study. Early in the term, when study rooms were limited, all students were strictly to observe the study period from 8:00 to 9:45 P.M. There would be no supervision of the period, "but the students will be required to satisfy the directors of their respective colleges that the period has been observed." No student was permitted outside the buildings or off the post during this period without special permission. The quarters and mess halls, used during the study periods, were to be kept orderly and quiet, and student-officers living in Beaune were also required to remain in their billets during the study periods, and "to satisfy the directors of their respective colleges that the period has been observed."[81]

Specific concerns were addressed as required, both those common to military bases in their normal course of operations, and those unique to Beaune, some revealing the difficulties of maintaining the spirit of a university in a military setting. The motor transport officer at Beaune, Captain L. A. Bonner, set up a regular schedule of motor cars to transport officers and civilian personnel—excluding enlisted men—on the short trip to Beaune or to the agricultural school at Allerey, with cars running every hour to Allerey and every half-hour to Beaune.[82] Trash disposal was a problem, and ashes were often dumped along the streets of the campus. Firewood and coal were not properly stacked or stored. Repeated instructions were required to get these matters rectified.[83] Many of the students sought to modify their living quarters in unofficial or unorthodox ways that had to be regulated. They were warned that it was prohibited to have stovepipes coming out of windows or the sides of buildings. Such installations were to be removed immediately and application made to the superintendent of buildings and grounds to have alterations made by qualified engineers. Other difficulties were caused by clogged sewers and drains. Paper and trash improperly disposed of caused much of the trouble and steps were to be taken to correct such abuses.[84]

A general conservation plan was also instituted. Safe water sources were difficult to maintain because the water supplied by the city of Beaune was declared unsafe and the university had to supply its own. It therefore had to be conserved.[85] Electricity had to be restricted as well, and it was noted that electric lights were often left burning during the day. Steps had to be taken to end such waste.[86] Stationery and office necessities were in short supply. Orders were issued to use only half-sheets of paper when these would suffice. Only the absolute minimum of carbon copies was to be made, and both sides of each sheet of paper were to be used where practicable. Finally, Reeves ordered that disciplinary action be taken against any person found using onionskin paper for toilet paper, a practice which "must be stopped at once."[87]

One of the clearest examples of the clash between the military mode and usual university practice appeared when it was brought to the president's attention that straw votes were being taken regarding the presidential elections forthcoming in 1920. "While it is desired to offer every facility to the teachers and students of the A.E.F. University in order that they may get the greatest good out of the courses offered," he asserted, nonetheless, it had to be remembered that, "after all," the university was a military organization bound by army regulations. These prohibited political discussions. Accordingly, straw votes, and mock political conventions and assemblies, were "expressly prohibited." Reeves was confident that "a resourceful teacher" could readily find a substitute for this form of instruction. Later, however, Professor Fogg, director of the College of Journalism, noted that his students in Argumentation (English 40) were "boiling to investigate and write (briefs and arguments) on such questions as the League of Nations, the Monroe Doctrine, and the admission of Germany to the League," among other subjects. He wanted to know if it was all right for students to include written argument "for an instructor's eye" only. He was told to "use good sense—that's all that's necessary."[88]

When confronted with the problem that many instructors were not dismissing their classes on time, causing considerable confusion, the obvious solution seemed to be an army one: bring in the bugle! It was ordered that five minutes before the end of each recitation, "First Call" would be sounded, signaling all instructors to begin preparations for the ending of their classes. At the notes of "Recall" sounded five minutes later, instructors would promptly dismiss their classes. "Assembly" would be blown at the beginning of each period.[89] The performer was bugler Roy Albright who, from 8:00 A.M. to 4:00 P.M., made a total of forty-five bugle calls, "from three directions," from his tower at the intersection of University and Johns Hopkins roads.

If there remained an element of uncertainty about the anomaly of a university with its liberal traditions and methods being conducted by an army steeped in a much more conservative tradition, the school at Beaune rapidly assumed the modus operandi and trappings of a university. Not all aspects of the military mode were anathema to the smooth functioning of an institution of higher learning. The school was expeditiously set in motion: classes started; related intellectual endeavors were soon underway; and provisions for the athletic side so familiar in American colleges and universities soon materialized. The band, soon created, remained primarily a military organization, though called the American E.F. University Band. On April 26, a headquarters company, commanded by Captain E. Huff of the quartermasters, and attached to the 12th Provisional Regiment, was set up. The new company was composed of band members transferred from their provisional regiments.[90]

One of the extracurricular developments that attracted some attention was the arranging, by a Committee of Excursions, of a series of Sunday outings for all university personnel. These were designed to be walking trips through the nearby countryside, bringing Americans into contact with picturesque, often historic, sites, towns and villages, and with the French.[91] The first venture, scheduled for March 16, involved a walk to the town of Savigny, about five kilometers from the university. Regimental kitchens furnished luncheons, and each man carried a canteen and his mess kit. Those with Kodaks were encouraged to bring them along, and art students reminded to take along paper and pencils for sketching, these being supplied to them for that purpose, as were maps for all participants. Points of interest included several picturesque châteaux, an old church, and wine cellars and stills.[92]

On March 23, the destination was the town of Meureault, about five miles away. As always, the emphasis was on informality, but all were reminded that they must not trespass, and were to respect the rights and customs of French citizens, thereby maintaining cordial Franco-American relations.[93]

The little town of Bouze was the destination for the next weekend, March 30, but inclement weather intervened, the trip being rescheduled for April 6.[94] By this time, the informal routine was well established, and apparently operated smoothly thereafter.

Special lectures and presentations became a familiar part of the university's calendar, and the students were subjected to the words of various and sundry politicians, economists, business leaders, "up-lifters," preachers, and other orators in a steady stream. Typical was one Saturday afternoon address by William Allen White, well-known journalist and author, and editor of the *Emporia [Kansas] Gazette*. Though his lecture, "Journalism," was primarily for journalism students, all personnel were invited. Other prominent journalists who were then covering the Paris Peace Conference were invited to address the journalism students, but their duties kept them tied close to Paris. After White, only one other appeared: Reginald Wright Kaufmann, of the *New York Tribune*, who was also a novelist. His topic was "The Newspaper and the Novel." The basis for much of his presentation was his own widely-acclaimed new novel, *Victorious*, the setting and theme of which was the American army in France. Kaufmann maintained that the newsman came to know life at many levels, which was "the best equipment for a novelist."[95] Another who spoke about literature was Henry Kitchell Webster, a short-story and novel author of Chicago who addressed the College of Journalism, and other interested students, on "Some Essentials of Fiction Writing."[96]

Another prominent speaker was Alfred C. Lane who delivered a general lecture on "Economic Stakes of the War." In a speech that had been delivered at various army posts in the United States, Lane led his audience "to the realization of the fact that the War is not over until the proper allocation of coal, iron and materials basic to Civilization has been provided [to all]."[97] The entertainment officer also saw no reason why lectures given by a Mr. Kline on "Electrical Improvements in Industries in the United States," would not draw interested audiences, and scheduled him for two evenings.[98]

Professor John Erskine, the university's educational director, presented two evening lectures on "Poetry and Poetic Experience."[99]

Another member of the university's staff who spoke was Captain Warren Abner Seavey, director of the College of Law, who lectured to the journalism students on the libel laws, during a week-long series of presentations.[100]

Other visiting orators included Lt. Gov. Mason S. Stone of Vermont, who was also the former educational commissioner of that state, an old crony of Reeves's, and "an educator of National reputation." Dr. Hugh Black, "one of America's most famous speakers," also addressed the students.[101] Black was a well-known Scottish-American educator, writer, and professor at the Union Theological Seminary in New York City. At a mass meeting of students and faculty at the university on the evening of May 14, he emphasized that the United States must give a new vision to the world, and that its citizens "must prepare to see life in terms of service and sacrifice." University men should strive to create a new aristocracy, not based on birth and privilege, but "by what they give and not by what they receive." Black was on the AEF lecture circuit that included Beaune.[102]

Much more significant was an address delivered by Colonel George Catlett Marshall of GHQ, Chaumont, speaking on the "Organization and Development of the A.E.F. and the Operations of its Army." Reeves himself, having read the text previously, warmly praised the address, noting that it was a most interesting one, and "it is believed will be one of the most instructive lectures which has been offered by the University."[103]

Marshall's presentation was the result of an initiative by General Pershing, who was strongly convinced that the French were trying to downplay the American contribution to the final victory. Pershing therefore intended that "as many as possible of the home-going A.E.F. be made acquainted with America's part in the war, from the larger viewpoint that it was possible for the General Staff only to have," of the progress of events and their bearing on the war's outcome.[104] More specifically—and at the same time to instill pride in the U.S. soldier—he selected nine hundred doughboys who in civilian life were editors, writers, or publishers and sent them on a special two-week tour of the AEF, so that when they returned to their civilian pursuits, they could tell the American people about the contributions of their troops in Europe. He also commissioned Hugh Drum and Willey Howell—in addition to Marshall—to make lecture tours of divisional camps to elaborate on the AEF story.[105]

After several aborted scheduled times, Marshall's lecture was finally set for May 26 and 27, at the University Theater at 7:45 P.M. Reeves pointedly suggested that all officers and instructors attend at least one of the lectures, with students being present by regiments, though regimental commanders were authorized to excuse their men, but "not to exceed 25 per cent of their strength."[106] Marshall related the course of movements and battles, using maps to illustrate his remarks, and also explained the strategical reasons behind operations. His story was "forcefully told," and was regarded as one of the university's stellar lecture presentations.[107]

Another compulsory lecture addressed that perennial concern of the AEF: venereal disease. Therefore, the lectures presented by an officer from the office of the chief surgeon, AEF, were more military than university business. The surgeon made eight presentations over a period of several days so that all personnel could attend.[108]

Within a month of the school's establishment, it was operating surprisingly efficiently and smoothly. Catching the spirit of the new enterprise, the registrar, Richard Watson Cooper, noted in early March that just three weeks previous, there had been "no college catalogs to serve as guides, no college faculties to consult, no courses announced, no college buildings in sight, and little save French mud out of which to make a great American University." Yet, almost overnight, one had been created, and he hoped that subsequently "our life together here upon these free fields of freedom-loving and art-loving Burgundy [will be] a distinct gain to the citizenry of our own country."[109]

SOLDIER-SCHOLARS

Notes

1. The Fifth Section of the General Staff, General Headquarters, AEF, was the section controlling all educational work in the American Expeditionary Forces. Head of the educational sub-section at G-5 for a time was Colonel M.A.W. Shockley; he was replaced by Rees.

2. Memorandum, to Dr. John Erskine, educational director, American E.F. University, and chairman, Army Educational Commission, from Colonel Ira L. Reeves, May 13, 1919, in folder "A.E.F. University, Historical Data," Entry 409, Box 1956; General Orders No. 1, Headquarters, American E.F. University, February 16, 1919, Entry 412, General Orders, 1919, based on Paragraph 110, Special Orders No. 44, GHQ, AEF, Chaumont, February 12, 1919. Reeves assumed command at the former Hospital Center at Beaune and all organizations and troops there, except for the Base Hospital Units still stationed there. See also General Order No. 30, GHQ, AEF, Chaumont, February 13, 1919, reproduced in U.S. Army. The American E.F. University, Beaune, Côte D'Or, France. Bulletin 91, *The Catalogue* (Dijon: Darantière Printer, 1919), part 1, pp. 5-8. Hereafter cited as *The Catalogue*, part 1, pp. 5-8. Part 2 of the *Catalogue* consisted of a listing of the names and addresses of the staff and students of the university, numbering over 10,500 names. It was printed in Dijon after the closing of the school and the return of the students and faculty to the United States. Reeves's title was later changed from superintendent to president. See General Order No. 20, Headquarters, American E.F. University, March 20, 1919, Entry 412, based on Special Order No. 77, GHQ, AEF, March 18, 1919.

There are several schematics and blueprints of the hospital complex at Beaune and the new construction to be undertaken in Box 1954, Entry 418. The projected new work was approved by Reeves on March 4, 1919. The supervising architect was Grosvenor Atterbury.

3. There is an article on Reeves in *A.E.F. University News*, vol. 1, no. 6, May 30, 1919. He was awarded the honorary degree of doctor of letters from Norwich University in 1916, and the honorary degree of doctor of law from Middlebury College in 1917.

4. Telegram, February 26, 1919, from Brig. Gen. Robert C. Davis, the Adjutant General, GHQ, AEF, Chaumont, France, to Commanding Officers of Divisions, AEF, in folder "A.E.F. University. Historical Data," Box 1956, Entry 409. Division commanders were to see to it that the soldier-students reported in full field equipment, including their arms, but without ammunition.

5. General Order No. 12, Headquarters, American E.F. University, March 1, 1919, Entry 412.

6. See numerous documents pertaining to this problem in folder "179: Instructors," Box 1925A, Entry 419.

7. Memorandum, from Col. Ira L. Reeves to Dr. John Erskine, Educational Director, American E.F. University, Headquarters, American E.F. University, February 27, 1919, folder "A.E.F. University. Historical Data," Box 1956, Entry 409.

8. Ibid.

9. See for example, General Orders Nos. 3 and 7, Headquarters, American E.F. University, February 16 and 22, 1919, Entry 412.

10. General Orders No. 9, Headquarters, American E.F. University, February 26, 1919, Entry 412.

11. General Orders No. 10, Headquarters, American E.F. University, February 27, 1919, Entry 412.

FOUNDING THE UNIVERSITY

12. See in Entry 412. Also this was in accordance with stipulations as defined in General Orders Nos. 9 and 30, issued by GHQ, AEF, at Chaumont.

13. See article on Erskine in *A.E.F. University News*, vol. 1, no. 6, May 30, 1919.

14. Copies of the Minutes of the University Council are in folder "College of Letters, A.E.F.U. Faculty Minutes, Orders, Memos," Box 1930, Entry 419.

15. Chief of the assistant directors was Captain John L. Gaunt, assistant to Dr. John Erskine, the educational director of the university. There is an article on the assistant directors in *A.E.F. University News*, vol. 1, no. 7, June 6, 1919. See also listing of all of the assistant directors in General Orders No. 21, American E.F. University, March 23, 1919, Entry 412.

16. See "Minutes of the First Regular Meeting of the Faculty of the College of Business," March 11, 1919, in folder "Minutes of Faculty Meetings," Box 1911, Entry 420.

17. See in Entry 412.

18. *The Catalogue*, part 1, pp. 16-19.

19. Memorandum from Reeves to Erskine, Beaune, February 27, 1919, in folder "A.E.F. University. Historical Data," Box 1956, Entry 409.

20. There is an article on Watrous in *A.E.F. University News*, vol. 1, no. 6, May 30, 1919.

21. See article on Hurley ibid. Patrick Jay Hurley later became the U.S. secretary of war in the Hoover administration (1929-1933). Recalled to active duty after Pearl Harbor, he became a brigadier general. In 1942, he became minister to New Zealand and, later in the war, was dispatched on a mission to Moscow. In November 1943, he was sent to China to prepare for Chiang Kai-shek's meeting with FDR and Churchill at Cairo. In August 1944, he returned to China as liaison officer with the Chinese Nationalist government, later becoming the U.S. ambassador to China. He returned to the U.S. in November of 1945.

22. See article on Dallam ibid.

23. Ibid.

24. See Ford discussed ibid., vol. 1, no. 4, May 15, 1919.

25. See article ibid., vol. 1. no. 6, May 30, 1919.

26. Ibid., vol. 1, no. 7, June 6, 1919. Dijon was characterized as "a students' mecca." There they viewed the famous Jacquemart clock, Notre Dame cathedral, a museum that ranked second only to the Louvre, the palace of the Burgundian dukes, and sights along the rue Liberté, as well as many other churches, squares, and gardens.

27. Memorandum No. 60, Headquarters, American E.F. University, May 20, 1919, Box 1944, Entry 415. Article in *A.E.F. University News*, vol. 1, no. 6, May 30, 1919.

28. See article in *A.E.F. University News*, vol. 1, no. 6, May 30, 1919.

29. There is a lengthy article detailing the school's organization in *A.E.F. University News*, vol. 1, no. 5, May 22, 1919.

30. Article in *Stars and Stripes*, May 16, 1919.

31. *The Catalogue*, part 1, pp. 30-32.

32. Article, *A.E.F. University News*, vol. 1, no. 4, May 15, 1919.

33. Ibid., vol. 1, no. 6, May 30, 1919.

34. General Orders No. 6, Headquarters, American E.F. University, February 23, 1919, Entry 412. The registrar of the university supervised the enrolling of all the students in the entire school system of the AEF. His first assistant was charged with the registering of students in the AEF University; the second assistant registered all students in the AEF's post and division schools; the third assistant registrar had charge of registering all AEF

students enrolled in French and British universities. See also General Orders No. 21, Headquarters, American E.F. University, March 23, 1919, Entry 412. This order lists the original directors of each of the colleges and their office locations.

35. Pertinent details of the school and faculty are in General Orders No. 11, Headquarters, American E.F. University, February 27, 1919, Entry 412. See also *The Catalogue*, part 1, pp. 57-58. See list of the successful candidates in memorandum, from the Adjutant General, A.E.F., to the Commandant, A.E.F. University, GHQ, AEF, May 13, 1919, and other documents pertaining to this subject in folder "Military Academy," Box 1923, Entry 408. There is an article on these men in *A.E.F. University News*, vol. 1, no. 5, May 22, 1919.

36. *The Catalogue*, part 1, pp. 59-60.

37. See article in *Stars and Stripes*, March 21, 1919.

38. Much of the following information is from Report of the College of Correspondence of the American E.F. University, J. Foster Hill, to John Erskine, Beaune June 1, 1919, in folder "College of Corr. A.E.F.U. Its Work and Organization," Box 1926, Entry 419. See also photographs of the staff ibid., folder "College of Corr. A.E.F.U.—Photographs"; and several folders containing syllabi and lesson plans, ibid. The assistant educational director was 1st Lieutenant Charles Burnell Fowler of the 139th Infantry. William Henry of New York supervised commercial courses and Dean Louis E. Reber of the University of Wisconsin headed engineering and industrial instruction. Elmer Warren Cavins, of the Illinois State Normal University, was the college's secretary and editor of publications. First Lieutenant Charles Burnell Fowler was the assistant director.

39. Hill reported that in addition to these regularly enrolled students, the college sent out books, outlines, and examination questions for their own use to 4,650 additional students after the college of correspondence had closed. No lessons could, of course, be received from them. He also reported that some 6,000 additional applications could not be filled because of a lack of books and materials. See ibid.

40. There are copies of these materials in Boxes 1927-1929, Entry 419. See reference in *Stars and Stripes*, March 28, 1919, and article in *A.E.F University News*, vol. 1, no. 6, May 30, 1919.

41. See application form from the College of Correspondence in Entry 408, Box 1871.

42. See articles in *Stars and Stripes*, March 21 and 28, and April 11, 1919.

43. From Huntsville, Texas, Branch was in the Headquarters Company of the 815th Pioneer Infantry. See his letter in Box 1868, Entry 408.

44. Church was attached to U.S. Camp Hospital #4. See his letter of March 21, 1919, in Box 1870, Entry 408.

45. Dickson was attached to the 56th Depot Company. See his letter of March 23, 1919, in Box 1874, Entry 408.

46. See Fenski's letter of April 1, 1919, in Box 1876, Entry 408.

47. See his letter in Box 1877, Entry 408.

48. See extensive report of the College of Correspondence of the American E.F. University, by J. Foster Hill to John Erskine, Beaune, June 1, 1919, in folder "College of Corr. A.E.F.U. Its Work and Organization," Box 1926, Entry 419.

49. See undated report on the courses in citizenship by 2nd Lieutenant Sedley Clarence Peck, of the Air Service, the assistant director, Courses in Citizenship, in folder "College of Citizenship A.E.F.U.—etc.," Box 1939, Entry 419.

50. See ibid.

51. U.S. Army. American E.F. University, Beaune, Côte D'Or, France. Bulletin No. 96, *Educate America. A Complete After-The-War Program For The Advancement of Public*

Education (Dijon: R. De Thorey, 1919).

52. U.S. Army. American E.F. University, Beaune, Côte D'Or, France. Bulletin No. 98, *The Community and the Home. A Preface to the Discussion of City Planning and Housing* (Dijon: R. De Thorey, 1919).

53. U.S. Army. American E.F. University, Beaune, Côte D'Or, France. Bulletin No. 101, *Some Industrial Problems* (Dijon: R. De Thorey, 1919).

54. U.S. Army. American E.F. University, Beaune, Côte D'Or, France, Bulletin No. 111, *A Few Elements of Preventive Medicine* (Dijon: R. De Thorey, 1919); Bulletin No. 92, *The Principles of Democratic Government* (Dijon: R.De Thorey, 1919).

55. There is a lengthy discussion of this program in *A.E.F University News*, vol. 1, no. 4, May 15, 1919.

56. *Stars and Stripes*, January 24, 1919.

57. Ibid., March 14, 1919.

58. *A.E.F. University News*, vol. 1, no. 7, June 6, 1919.

59. See undated report on post schools, by Major L. Frazer Banks, in folder, "Reports—Post and Division Sch.," Box 1930, Entry 419; undated and unsigned report on both Post and Division schools, ibid. These reports were apparently completed following the end of the school term, on or about June 1, 1919, when so many other of the reports of the various colleges were drafted.

60. See unsigned, undated report on the Division School, and unsigned, undated report on both the Post and Division schools, in folder "Reports—Post and Division Sch.," Box 1930, Entry 419. Both of these reports were obviously drafted following the end of their respective courses, probably on or about June 1, 1919. See also *The Catalogue*, part 1, pp. 122-27.

61. General Orders Nos. 4 and 9, Headquarters, American E.F. University, February 20 and 26, 1919, Entry 412.

62. Details are in General Orders No. 4, Headquarters, American E.F. University, February 20, 1919, Entry 412.

63. See in General Order No. 8, Headquarters, American E.F. University, February 25, 1919, Entry 412.

64. Among these were large mechanical and electrical engineering laboratories with three big hangers being erected to house generators, meters, and motors, as well as gas engines, boilers, three locomotives, five airplanes, and road making equipment. See article in *Stars and Stripes*, March 28, 1919.

65. For details of these workers, see ibid., February 21 and May 16, 1919.

66. See in Memorandum No. 4, Headquarters, American E.F. University, March 9, 1919, Entry 415.

67. Memorandum No. 11, Headquarters, American E.F. University, March 19, 1919, Entry 415.

68. General Orders No. 24, Headquarters, American E.F. University, March 28, 1919, Entry 412.

69. Bulletin No. 46, American E.F. University, April 2, 1919, Box 1944, Entry 414.

70. General Orders No. 28, Headquarters, American E.F. University, April 6, 1919, Entry 412.

71. Memorandum No. 8, Headquarters, American E.F. University, March 16, 1919, Entry 415. See undated memorandum, Captain Sherman Child, to Major Babcock, in folder "Salutes," Box 1923, Entry 408, which became the basis for Memorandum No. 8.

72. See General Orders No. 13 and 18, Headquarters, American E.F. University,

March 1 and 12, 1919, Entry 412. See large, bound notebook used for signing out for passes and leaves, in Box 1964A, Entry 420.

73. Donald Smythe, "Venereal Disease: The AEF's Experience," *Prologue*, 9 (Summer 1977): 64-74.

74. Memorandum No. 6, Headquarters, American E.F. University, March 13, 1919, Entry 415, and Bulletin No. 53, Headquarters, American E.F. University, April 8, 1919, Entry 414.

75. See records of several summary courts martial in Box 1877, Entry 408.

76. See several loose sheets, "History of the Medical Department of the American E.F. University," in Box 1964A, Entry 420.

77. See daily reports on these people in folder "Daily Reports of A.E.F. Univ., Personnel Residing in Beaune," Box 1906, Entry 408; Memorandum No. 5, Headquarters, American E.F. University, March 10, 1919, Entry 415.

78. Memorandum No. 6, Headquarters, American E.F. University, March 13, 1919, Entry 415.

79. Bulletin No. 39, Headquarters, American E.F. University, March 27, 1919, Entry 414.

80. Bulletin No. 41, Headquarters, American E.F. University, March 29, 1919, Entry 414.

81. Memorandum No. 15, Headquarters, American E.F. University, March 25, 1919, Entry 415.

82. General Orders No. 29, Headquarters, American E.F. University, April 9, 1919, Entry 412.

83. Memoranda Nos. 16, 22, and 23, Headquarters, American E.F. University, March 26, and April 1 and 2, 1919, Entry 415.

84. Memoranda No. 10, 17, and 23, Headquarters, American E.F. University, March 18 and 26, and April 2, 1919, Entry 415.

85. Memorandum No. 25, Headquarters, American E.F. University, April 6, 1919, Entry 415; Bulletin No. 45, Headquarters, American E.F. University, March 31, 1919, Entry 414.

86. Memorandum No. 12, Headquarters, American E.F. University, March 20, 1919, Entry 415.

87. Memorandum No. 14, Headquarters, American E.F. University, March 22, 1919, Entry 415.

88. Memorandum from Prof. Miller Moore Fogg, director of the College of Journalism, to John Erskine, March 28, 1919, in folder "College of Journalism—A.E.F.U.—Correspondence to Educational Director," Box 1931, Entry 419. See also Memorandum No. 45, Headquarters, American E.F. University, May 1, 1919, Entry 415.

89. Memorandum No. 31, Headquarters, American E.F. University, April 12, 1919, Entry 415. *Stars and Stripes*, May 16, 1919; *A.E.F University News*, vol. 1, no. 4, May 15, 1919.

90. General Orders No. 36, Headquarters, A.E.F University, April 25, 1919, Entry 412.

91. Memorandum No. 6, Headquarters, A.E.F University, March 13, 1919, Entry 415.

92. Memorandum No. 7, Headquarters, A.E.F University, March 14, 1919, Entry 415. These parties departed Beaune at 8:30 A.M., and were expected to return by 5:00 P.M.

93. Memoranda Nos. 6 and 13, Headquarters, A.E.F University, March 13 and 21, 1919, Entry 415.

FOUNDING THE UNIVERSITY

94. Memoranda Nos. 17 and 24, Headquarters, A.E.F University, March 27, and April 5, 1919, Entry 415.

95. Article on Kaufmann in *A.E.F University News*, vol. 1, no. 4, May 15, 1919; and memorandum from M. M. Fogg, director of the College of Journalism, to the educational director of the University, May 31, 1919, in folder "Memos, Bulletins, Special Orders, etc.," Box 1932, Entry 419. This memorandum contained a summary report on the College of Journalism's work during the academic term. See also Bulletins Nos. 40 and 87, Headquarters, A.E.F University, March 28, and May 14, 1919, Entry 414, and *Stars and Stripes*, May 16, 1919.

96. *A.E.F University News*, vol. 1, no. 4, May 15, 1919.

97. Bulletin No. 43, Headquarters, A.E.F University, March 30, 1919, Entry 414.

98. Bulletin No. 30, Headquarters, A.E.F University, March 22, 1919, Entry 414.

99. Bulletin No. 81, Headquarters, A.E.F University, May 7, 1919, Entry 414.

100. *A.E.F University News*, vol. 1, no. 4, May 15, 1919. Seavey, in civilian life, was a professor of law at the University of Indiana.

101. Bulletins Nos. 85 and 86, Headquarters, A.E.F University, May 12 and 14, 1919, Entry 414.

102. Article on Black in *A.E.F University News*, vol. 1, no. 4, May 15, 1919.

103. Bulletin No. 103, Headquarters, A.E.F University, May 22, 1919, Entry 414.

104. *A.E.F University News*, vol. 1, no. 6, May 30, 1919.

105. Discussion in Donald Smythe, *Pershing: General of the Armies* (Bloomington: Indiana University Press, 1986), p. 247.

106. Bulletins Nos. 106, 107, and 109, Headquarters, A.E.F University, May 24 and 26, 1919, Entry 414.

107. Account in *A.E.F University News*, vol. 1, no. 6, May 30, 1919.

108. Memorandum No. 43, Headquarters, A.E.F University, April 28, 1919, Entry 415. These were in late April and early May.

109. For Cooper's remarks, see *Society As A University*, by John Erskine, published by the university as Bulletin No. 18, with attached note, "The Registrar Speaks," 12.

CHAPTER 3:
Setting up the Traditional Academic Colleges

THOUGH THE AEF university established several specialized colleges, such as the Cadet College and the Correspondence College, the traditional academic colleges familiar in America were at the heart of the school's efforts. One of the most active was the College of Business. But it seemed necessary to explain and justify its very existence within a university setting. Those in charge were aware that the term "Business College" was appropriate "perhaps 50 years ago [for] the institutions that aimed to train stenographers and bookkeepers for business concerns." Indeed, "as late as 15 years ago the idea of giving work of a university or college grade for business men was new and didn't mean very much." It was generally believed in the business community "that the only way in which a man could make a success out of business life was to jump into the game and learn by experience." While this notion had merit, it was by then recognized that, though no one could master business matters by study alone, nevertheless it could enable a businessman to "cut down by many years the period of apprenticeship and to equip himself to go forward much more rapidly than would otherwise be possible." Fifteen years prior to 1919, there were only two schools of commerce in the United States of university grade; by 1919, there were nearly thirty, with numerous others springing up in Europe and Asia. Therefore, the College of Business at Beaune aimed "to take men who have had either a good academic training or a good line of business experience and give them a preparation that will help them to step into executive or semi-executive jobs in business." In addition, the college promised that its students would be "undertaking work of the same high grade as that in the College of Letters, the College of Engineering, the College of Law or any other department of this A.E.F. University." Finally, the aim of the business faculty was "to work *with* you not *on* you," as befitted a staff that saw itself as "business men first and professors afterward."[1]

The man mainly responsible for launching the college was Dr.

TRADITIONAL ACADEMIC COLLEGES

William Henry Lough, of the faculty of the School of Commerce at New York University. He was director until April 28, when he was replaced by Major Edward Helsley of San Francisco. In civilian life, Helsley was in the nursery and land development business in California and Arizona. Captain Herbert Edmund Fleischner of Boston, formerly associated with the Industrial Services and Equipment Company, was the assistant director.

The work of instruction at the college was done by a faculty of sixty-six under the seven heads of departments. The courses with the largest enrollments were: Business Organization, with 1,187 enrolled; Principles of Selling and Advertising, with 1,024 students; and Commercial Law, with 990 students. Other popular courses included Principles of Accounting, Foreign Trade, and Insurance Practice and Salesmanship.

The courses and departments were under the direction of men with practical experience in the business world, or from the staffs of business schools and colleges. For example, the course in Business Organization was directed by 1st Lieut. Robert E. Geary, earlier an industrial engineer for the Bethlehem Steel Company; the course in Insurance Practice and Salesmanship was under the supervision of 2nd Lieut. Edmund Quincy Abbott, formerly a special agent for the Mutual Benefit Insurance Company; and 1st Lieut. Armistead E. Thames, who, before he entered the army, was manager of the Gulf Continental Steamship Company, conducted the course in Foreign Trade.[2]

The College of Business found that one significant way of reaching students and involving them in a practical way in business instruction was to organize specialty clubs, such as a Drug Club, Men's Clothing Club, Jewelry Club, Silk Salesmanship Club, China Club, Auto Club, and even a Bric-a-Brac Club. There were eventually sixteen of these organizations. In addition to regular meetings (usually in the evenings) to study retailing, wholesaling and salesmanship along their respective lines, club members often took tours of French business establishments and factories so as better to understand their respective fields. The Silk Salesmanship Club, for example, investigated the operations of the extensive French silk works in Lyon on April 26-27, while the China Club and Bric-a-Brac Club viewed the Haviland ceramic works at Limoges, and the Jewelry Club visited one of the world's largest watch factories in Besançon.[3] The clubs also met socially, as for example, when the twenty-five members

4. Auditorium under Construction, Beaune.   111-SC-161042

5. Campus in the Mud and Pershing Athletic Field, Beaune. 111-SC-161041

TRADITIONAL ACADEMIC COLLEGES

6. New Construction, AEF University, Beaune.   111-SC-161038

7. Registration Lines, Beaune.   111-SC-161030

of the Lumber Club held a banquet at the Hôtel de la Poste, in Beaune, on the evening of May 12.⁴

Surely one of the most ambitious of the university's undertakings was the Medical College, headed by Colonel Joseph Herbert Ford of the Medical Corps.⁵ Associated under the college, were the Colleges of Dentistry and Veterinary Medicine, and later the College of Pharmacy. First Lieut. Horace K. Alexander was appointed registrar for all the associated colleges and Capt. Thomas Stephen Brown became the assistant director of the Medical College. Lt. Col. Ferdinand Schmitter was made director of laboratories. It was initially decided that a four-year schedule would be drawn up with requirements equal to those of a Class A medical college in the United States. However, these plans proved too ambitious. In the event, it was determined that the second through fourth years of medical work could not be undertaken at Beaune, and that students at these levels, and those desiring graduate work, would be transferred, to French universities, more often to the medical schools at Lyon, Paris, Marseilles, Bordeaux, and Montpellier, or to British universities and colleges. Students desiring a postgraduate course in laboratory methods and technique were to be accommodated at the French Central Laboratory at Dijon, or the Pasteur Institute in Paris. Also plans for a pre-medical college were canceled, and pre-med students were accommodated within the College of Science for their basic science courses.⁶ For those qualified to begin work at Beaune, a course of study of thirty-five hours per week was set up, with lectures in anatomy, physiology, organic and physiological chemistry, histology, bacteriology and clinical microscopy. Laboratory courses were in the subjects of bacteriology, clinical microscopy, and pathology. Registration data revealed that numerous students desired to study pharmacy; consequently, a college of pharmacy was established under Captain Eugene E. Preston, an infantry officer who held a Ph.G. degree.

The medical program was initially handicapped by a shortage of laboratory equipment. Accordingly, the laboratories of the base's Camp Hospital No. 107 were for a time pressed into service as teaching labs. However, much equipment that had been left behind with the closing of the hospital complex, which had been in operation at Beaune before the end of the war, was located, and soon fifteen laboratories were in full operation, with the students themselves making and equipping their own labs. While lacking in sophisticated apparatus, the students "were taught that elaborate equipment was not necessary as long as they had self-reliance." Additional equipment and supplies came in from the AEF's own laboratory center, from French schools, such as the *Ecole de Médicine* and the Pasteur Institute at Paris, and from the French laboratory center in Dijon.[7] It was necessary to find supplies of cadavers and bones, both of which were also located. The first departments in full operation were those of pathology and bacteriology, which were joined by others as rapidly as possible.

The Dental College was directed by Lt. Col. Otis H. McDonald. The faculty consisted of ten commissioned dental officers, who were also on the staff of the dental infirmary on the post. The college emphasized practical dental work, the infirmary often being the site of instruction.[8]

The College of Veterinary Medicine was headed by Major George B. McKillip. This school also planned to organize a four-year program, but actually only offered first-year work. Advanced students were transferred to foreign schools, usually to the University of Edinburgh.[9]

The College of Pharmacy was interwoven with the other colleges. The courses were identical with those in the other schools, and no special classes were offered.

The entire College of Medicine maintained close ties with the College of Science where many of the medical students were enrolled. The total enrollment in the medical schools was 207.[10]

The College of Fine and Applied Arts, headed by George Sidney Hellman, with its enrollment of 420 students, consisted of five departments: architecture, painting, sculpture, commercial and industrial arts, and another devoted to field work. Indeed, much of the study by art students was done on field trips, some lasting for several days, but also much decoration and design work was done to beautify the university. Painters decorated club rooms and mess halls; artists designed the Army Educational Corps's insignia and the university flag; while one of the students designed and painted the curtain for the University Theater. Members of the classes in industrial and commercial art created over 150 posters for use by the army, such as those in a series on social hygiene. Sculptors at Beaune and at Bellevue, the special art school near Paris, entered a competition for the design of a tablet to be presented to the French universities where American soldier-students were studying. In addition, architects and sculptors at Beaune designed a monument to be presented to the citizens of Beaune in memory of French and

American war dead. The director of the college was George S. Hellman; the associate director was Lloyd Warren, who was also in charge of the school at Bellevue. The faculty, numbering more than thirty, included many noted artists, sculptors and architects including the American sculptor, Lorado Taft, who headed the department of sculpture; Prof. John Galen Howard, director of the School of Architecture at the University of California; Eleazer Bartlett Homer, former dean of the School of Architecture of Providence, Rhode Island, and head of the department of architecture; and the distinguished French architect, Jean Hebrard. Late in the school term, the students of the college displayed their drawings, watercolors and paintings in two buildings that were transformed into art museums for the occasion. Also, the better to inform the university about art matters, various evening lectures on art were given under the auspices of the College.[11]

One of the university's most substantial schools was the College of Engineering, headed by Dr. Louis E. Reber of the Army Educational Corps.[12] His assistant was Major Charles Beecher Stanton, of the 15th Engineers. The faculty consisted of a total of one hundred instructors, six of whom were from the Army Educational Corps, together with fifty-eight officers and thirty-six enlisted men. These taught virtually every subject of engineering science found in American colleges and universities of the day, with the exception of a few courses in mining engineering, notably those concerned with laboratory work in ore dressing and metallurgy. These were not offered because of the lack of laboratory equipment and the time necessary to obtain it. However, lecture courses in ore dressing and metallurgy were taught. The college occupied a large complex of buildings, seven that were originally hospital wards, and four large hangers.

The college's equipment came in large part from the huge stores of the AEF, especially from various engineer depots, and Signal Corps equipment stores, while the Air Service furnished planes and motors for the aircraft mechanics course.[13] The aviation department took pride in being the first of its kind to be permanently established "as an institution of education in connection with a university." Under the direction of 2nd Lieut. John Earl Black, of the 172d Aero Squadron, it enrolled thirty students. These had seven aircraft for use in practical training: a Liberty-powered de Haviland DH-4; two British planes, an Avro and a Camel; a French observation aircraft, a Salmson; two French Spads;

and a French Nieuport. However, no flight instruction was given, the emphasis being on aircraft construction, repair, and maintenance.[14]

The Department of Mineralogy and Mines took advantage of the mines and quarries in the vicinity of the university for practical instruction. In addition, the mineralogy collections of the curator of the museum in Beaune, as well as those of the University of Dijon, were placed at the disposal of the college. Permission was secured to use an approximately 1½-by-2½-mile tract of land that was a few miles north of Beaune and included two small villages, for surveying classes.

The college included the departments of civil, electrical, mechanical, and mining engineering, offering a total of sixty-three courses, with a final number of 2,556 enrollments. In addition, 912 students enrolled in vocational training courses given by the college for the post and division schools. The College of Engineering was one of the soundest and best of the colleges and was efficiently run and managed.[15]

The Agricultural College at Beaune was under the direction of Dr. Kenyon Leech Butterfield, president of the Massachusetts Agricultural College, and a member of the Army Educational Commission. While much of the university's agricultural endeavor focused on the separate Farm School at Allerey, there was sufficient demand for advanced studies in agriculture for a full-fledged college of agriculture to be established on the campus at Beaune. It enrolled about 676 students, taught by a faculty of 59. The dean of the college was Harry Hayward, dean of the Delaware Agricultural College. The assistant director was Capt. Edward N. Wentworth, former professor of animal breeding at Kansas Agricultural College. Four departments—animal husbandry, agronomy, horticulture and forestry, and rural economy and sociology—offered over forty courses. Hayward observed that the attendance had been outstanding "and the spirit and enthusiasm shown in the work has been so marked that it has frequently caused favorable comment from those educators who have had an opportunity to observe the character and quality of the instruction in this College."

The army supplied the college with its supplies and equipment, including several tractors, ranging from large caterpillar types to smaller machines, including one steam tractor. The college was able to offer most of the classes requested by students, with the exception of landscape gardening.[16]

SOLDIER-SCHOLARS

The first college to open its doors to the students—by a few days—was the College of Law. Its director was Captain Warren Abner Seavey, in civilian life professor of law at the University of Indiana. The faculty and administrative staff consisted of four officers, two noncommissioned officers, and one member of the Army Educational Corps. The staff was supported by the enthusiastic law students, closely bound together by their common interests. They were perhaps the most cohesive group on campus. They helped keep their rooms clean, did much of the clerical work, and in other ways assisted in the school's functioning. They erected and furnished their first classroom as well. The faculty taught eleven courses to about 150 students. Obtaining textbooks was a problem that was only later alleviated, most notably with the arrival of key books and sources from England. The college emphasized the casebook method of study, based on the Socratic dialectic approach, and in the evenings conducted moot courts, which aided the college to reach its goals of assisting its students "to think in [a] lawyer-like manner and to understand a few of the law's mysteries." Indeed, though rather small and with a minimum of books and supplies, the faculty insisted that it had all that was necessary to make an effective law school, and support "a student body with an intense desire to learn and a group of older brothers, somewhat better acquainted with the law, to inspire the vision of service."[17]

The College of Science utilized seventy instructors who taught forty-seven courses, in eleven departments, to the 820 students enrolled, with a total course registration of 2,358. Its director was Lt. Col. William Freeman Snow, who also headed the Citizenship Course. The largest registration was 1,003 in mathematics; the smallest—in astronomy—was five. The school required a large number of laboratories, and drawing upon the resources available in the army depots of the Medical Department, Chemical Warfare Service, and the Quartermaster's Department, with textbooks supplied by many sources in the AEF, the students were reasonably well provided for. For each student in the biological laboratories, for instance, all the essential equipment and supplies, including a microscope with oil immersion lenses, and other accessories, were available. Each laboratory was provided with sterilizers, incubators, microtomes, and other apparatus. The chemical laboratories were similarly equipped, and each student's desk was provided with the usual reagents and glassware in addition to such appliances as acetylene burners, water and compressed air connections, sinks, hoods, water baths, crucibles, stills, analytical balances, and standardized volumetric instruments. In the teaching of geology and public health, extensive field work was conducted, supplementing the classroom and laboratory instruction. The college recognized that its students were readily broken down into three groups whose needs needed to be anticipated: those who required instruction in math or training in a laboratory science in order to support their main programs in engineering, medicine, education, agriculture, or other professional programs; those who desired to obtain instruction and practical training so as to prepare them to function as technicians or inspectors in laboratories and industrial establishments; and those who simply wished instruction in the courses offered for their general cultural value. Initially, it proved difficult to obtain a competent faculty, and many of the men who reported as instructors in chemistry, for example, were "unprepared to take charge of the courses which are to be given." Therefore, the scientists recognized that "if the work of the University in this Department is to be of a grade equal to that of the work in the average American College, it will be necessary to secure the services of the men who have had experience in College teaching." Accordingly, the army's Chemical Warfare Service was ordered to detail to Beaune qualified personnel who had both the training and experience to enable them to function as instructors at the college level. These men were shortly on their way, and the college was soon adequately staffed.[18]

The College of Journalism was dedicated to the goals of relieving "the beginner of most of the miseries that beset a 'cub' reporter in his first year's work," to train him in the work of an editor, and instruct him "all about a newspaper plant; how it is organized and how it functions." The college also placed emphasis on the finer points of journalistic writing. In addition, considerable weight was placed on the social aspects of the work of the journalist, laying a solid foundation in those subjects essential to the broadly-equipped newsman: economics, history, literature, philosophy, psychology, political science, and science. This ambitious undertaking was in the hands of Prof. Miller Moore Fogg, the director of the college. Fogg was a graduate of Brown University in 1894; he obtained the A.M. there the following year, and did additional graduate work at Harvard in 1901. For some time, he was manager of the news bureau of the *Daily Press* in Asbury Park, New Jersey. He later served as correspondent and special writer for the *New York Sun*, the *New York Evening Post*, the *Boston Transcript*, and the Associated Press. He then went to the University of Nebraska, where for twenty years, as professor of rhetoric, he was in charge of the university's courses in journalism. The college's assistant director was Captain Archie Keefer Rupert, of the 137th Infantry. He possessed an A.B. from the University of Indiana and had worked on the *Kansas City Star* and the *Kansas City Journal*. There were seventeen instructors, fourteen of whom, all with practical experience in the newspaper field, were from the military. Of these, all had attended college, seven of them holding degrees. Of the latter, two were from the Marine Corps: Sergeant Major Guy Douglas Wilson of the 6th Regiment, who had his degree from Baylor University and was news editor on the *Waco Morning News*, and later on the *Fort Worth Record*; and Private 1st Class Herman J. Mankiewicz, 5th Regiment, whose A.B. was from Columbia University. He had worked on the *New York Tribune*, and was later managing editor of the *American Jewish Chronicle* in New York City. In addition to the instructors from the military, there were three others who were civilian professors on leave from the University of Chicago, Iowa State University, and the University of Nebraska.[19] The 137 journalism majors produced 523 registrations in the seven courses taught, which included The Special Article, Newspaper Editing, The Editorial, and Agricultural Journalism. The school also drew upon advertising courses taught by the department of business English in the College of Letters further to enhance the budding journalists' skills in this important aspect of newspaper work. In addition, "a concession to fiction" was made in a short story course, in which the building of plots and writing of dramatic narrative was taught. The college was unable to meet the requests from students for a course in cartooning, and because of the lack of a fully-equipped printing plant, it was impossible to give instruction in printing, which at least two dozen students desired.[20]

One of the most practical means of instruction was a campus newspaper, which the University Council recommended and Colonel Reeves approved.[21] For three issues, the weekly was an eight-page mimeographed publication, but by an agreement with the Darantière press in Dijon, the last four issues were published as a six-columned, four-page regular newspaper, each with a press run of five thousand copies. In addition to being a means of instruction for the students, the paper was intended to help with public relations for the university. To these ends, copies of the *News* were sent to leading newspapers, and most of the college and university libraries, as well as to numerous public libraries in the United States. The first issues were styled the *Bulletin of The A.E.F. University News*; the last four, the *A.E.F. University News*. The staff was composed of both officers and enlisted men, the managing editor for most of the time being 1st Lieut. Marcus E. Sperry. These sheets, rich sources for reconstructing much of the ambience of the university, also "formed a feature of. . .campus life greatly appreciated by the student body."[22]

Also, so as to enhance the instruction given by the College of Journalism, eight well-known journalists and authors, who were then attending the Paris Peace Conference, were engaged to lecture at the university. However, only William Allen White of Kansas, who delivered an address on "Reporting the Peace Conference," and Reginald Wright Kaufmann, who spoke on "Journalism and the Novel," participated. The early closing of the university and the inability of others to get away from their duties at Paris prevented their participation.[23]

Dr. Joseph M. Gwinn, director of the College of Education, had been superintendent of schools at New Orleans before coming to France. The college's assistant director was 1st Lieut. Louis John Schmerber, formerly superintendent of schools at Compton Lakes, New Jersey. Other members of the faculty, numbering twenty, were college and university faculty or public school officials. Only two were from the army, the others being civilians. Typical of the civilian staff were Mark E. Penney, who was dean of the Teachers' College, Syracuse University, and Elbert Waller, superintendent of schools in Grayville, Illinois. The college, which enrolled 232 students in the fourteen courses that it taught, sought to build upon mature students who, as soldiers in the war, already possessed a broader view of the world, a fuller appreciation of mankind, and a better preparation for entering the teaching profession following their return home. Because the faculty was a cosmopolitan group, no provincial methods, customs, or theories were entrenched in the college, and the many educational points-of-view and practices represented were given due consideration, to the benefit of all.[24]

Beyond its usual instructional functions, the college discharged numerous other duties. William Everett Stark, in addition to heading the department of elementary school management, was the supervisor of instructors in post and divisional schools throughout the AEF. A course in "Better Americans for America," designed and delivered in the college, sought to instruct teachers in the methods of teaching adult foreigners desiring to obtain American citizenship, emphasizing not only the three "R's," but also better means "of making NEW citizens BETTER citizens," for the nation. To further enhance their knowledge of educational methods and theories of French educators, field trips to French schools frequently supplemented classroom instruction, as did observation of instruction in post and division schools, especially those connected to the College of Education on the campus at Beaune.

For a two-week period beginning on April 14, 494 soldier-teachers from the AEF's system of post schools arrived for a special intensive training program. This was intended to arouse their educational enthusiasm, to give them some instruction, and to send them back to the combat divisions better able to instruct their comrades-in-arms. The method was to immerse each student in one course, selected from a list of eighteen, for the entire period. Of those ordered to Beaune for this work, not all were educationally qualified, and it was impossible to certify all who were sent. Indeed, incredibly, three of the students were illiterate, while numerous others had only finished the fourth grade. Some 73 had only a "slight experience in high school," while 204 had some work in college and 65 were college graduates. At the end of the session, the college certified 385 of the 494 as qualified to assume their duties as instructors in the post and division schools throughout the AEF.

The children of Beaune also derived benefit from instruction in physical education conducted by teachers supervised by the college and under the direction of Major W. N. Howard, the university's athletic officer.[25] In addition, on five nights a week at the *Lycée de Beaune*, the college taught classes in conversational English for various age groups, for the benefit of three hundred citizens of Beaune. Also, each afternoon for an hour, several English classes were conducted on campus for the benefit of the French employees of the school.[26]

One of the few military heads of a college was Major William Hammond Parker of the College of Letters.[27] It was the largest on campus, enrolling 3,100 students in 148 classes. In civilian life, the major was a professor of political and social science at the University of Cincinnati. His assistant was Captain Roscoe Edward Parker, formerly of the state educational department of North Carolina. Major Parker succeeded Dr. Herbert D. Foster, who had organized the school, but who took charge of the College of Science when this was separated from the College of Letters. Much instruction in the college took the form of field trips to various locations in France. In addition, the college organized a debating club that challenged members of other clubs organized by soldier contingents at French universities, such as at Grenoble, Lyons, and the Sorbonne. The college maintained nine departments: English, economics and social science, history, political science, philosophy, Latin, Greek, French, and Spanish. The Department of French was the largest in the college. Included in its faculty of forty-seven were twenty-five French soldiers comprising the French Mission militaire française. These were all winners of a *croix de guerre* and several had won, in addition, the *médaille militaire* or *Légion d'honneur*.[28] One thousand three hundred and ten students were enrolled in French. The citizens of Beaune organized the Franco-American Club in the city expressly for members of this department. The Department of Spanish had been added in response to student demand and had a class registration of 500.[29]

The College of Music was set up with the idea that music students would encounter a syllabus similar to that of a conservatory of music.[30] There were 748 registrations for all the courses, divided as follows: 436 piano students, 78 violin students, 18 cello students, 40 voice students, and 176 students of band instruments. In addition to classes in their specialties, all of the students were required to take a theory course, some taking an optional course in music history. The college was able, after some effort, to procure a qualified faculty, which numbered thirty-one.

The major difficulty that the College of Music faced was that there were almost no pianos or stringed instruments available, though there were adequate numbers of band instruments. Only late in the session were enough pianos and stringed instruments procured for classes to be conducted in these areas. Consequently, many faculty members were idled, unable to contribute to the program as they could have done. This led to considerable frustration. The college was accordingly able to function adequately only in its theory, band instrument, history, and voice departments.

The college was headed by Franklin Whitman Robinson, of the Army Educational Corps. He possessed an undergraduate degree and an A.M. from Columbia University. He was head of the department of aural theory of the Institute of Musical Art in New York City. The assistant director was First Lieutenant Anselmo Fulton Dappert, of the Tank Corps, in civilian life a civil engineer. It is interesting to note the origin of some of the faculty members, what they were in the army and what they had been in civilian life. Private 1st Class Wilson Poitevent Fraser, of the 114th Infantry, was an instructor of piano in a conservatory of music in Sherman, Texas, before entering the army. Private Clarence S. Haines, of the 113rd Engineers, was a private teacher of voice from Indianapolis, Indiana. Second Lieutenant Hypolite Theophele Landry, of the 141st Field Artillery, was a private teacher of trumpet and piano in New Orleans, Louisiana. Private John Kunkelman Miller, of the 64th Infantry, was a private teacher of violin in Port Arthur, Texas. One of the most active of the faculty members was Walter Squire, of the Army Educational Corps, who held an A.M. from Northwestern, and other graduate degrees in music from the University of Washington, where he was head of the departments of piano and theory and harmony before coming to France.

The man who led the AEF University Band was Private Vincent Earl Truxell, of the 315th Machine Gun Battalion. In civilian life he was director of the Archer School of Modern Piano Playing in Pittsburgh, Pennsylvania. He established an excellent band at Beaune, which was perhaps "the most successful result of the activities of the College of Music." The individual members were thoroughly drilled and the students benefitted from individual instruction as well as from the study of musical theory. The band frequently performed at official functions, and staged numerous concerts for the benefit of the citizens of Beaune.[31]

The theory department also functioned successfully, the class in notation being conducted by Private 1st Class Frederick E. Bastel, of the 101st Infantry. In civilian life, he was an instructor of music in the Oberlin, Ohio, high school. Harmony courses were taught by Squire and by Private 1st Class Edward French Hearn, of the 323rd Field Artillery, formerly an instructor in theory and harmony at Westminster College, New Wilmington, Pennsylvania. Squire also taught successful courses in music history.[32]

The voice department was under the direction of Wilford Watters, of the Army Educational Corps, who held a B.M. degree from the National Conservatory of Music. He was on leave from his post as director of the voice department of the Atlanta Conservatory of Music, Atlanta, Georgia.[33]

Though the College of Music functioned as well as it could under the circumstances, its director was of the opinion that its greatest accomplishment was in helping to change the attitude of many soldiers on the eve of going home. Many doughboys had discovered that the American Army did not stand for destruction alone; the health and welfare of the minds of the soldiers were considered and looked after as much as the health and welfare of their bodies. Most striking had also been "the intense desire of the [individual soldier] to avail himself of the educational opportunities offered him and his enthusiasm in taking up the work laid out for him," which would "never be forgotten by the men who were privileged to serve him." Equally memorable was "the manner in which the music students who happened to be in the College building fell upon and tore open the first crates of instruments that arrived." Many of the men had not seen a musical instrument "in all of those dreary days at the front, and at last they were able to follow up their hearts' desires and work towards things constructive in their lives." Robinson "sincerely hoped" that the College of Music "has added its part in the sum total of influence that the A.E.F. University at Beaune will be determined as having presented to the life of the Army." Not least, he hoped that the college had made some progress in raising the standard of musical activity in the army of the United States.[34]

One of the innovations at Beaune was the use of enlisted instructors who often taught classes in which officers, up to the rank of major, were enrolled. While it seemed strange for a time, instructors and students alike became accustomed to the novelty, and it was soon accepted practice. The instructors were recognized as being in charge, and no clashes were reported. These enlisted men were drawn from all ranks and from all branches of service, including the Marines. In civilian life, some of the men had held professorships in American universities; others had been executives in firms and businesses; yet others were attorneys, artists, painters, sculptors, musicians, editors, and journalists. The enlisted instructors, numbering more than two hundred men, were organized separately into the 4th Provisional Regiment and had their own billeting area, clubroom, athletic field, and mess.[35]

While the vast majority of the personnel in the AEF University were men, there were a substantial number of women involved, though not as formally-enrolled students. There was a contingent of army nurses stationed at Beaune, and women members of the Army Educational Corps. Other women were prominent in work of the YMCA, especially those who staffed the university's regimental huts. There were five of these buildings, involving seventy-two women engaged in canteen and educational work, as well as in managing the events on the social calendar, which emphasized dances. One woman computed that on average she covered 12,000 yards a week—almost seven miles—on the dance floor, "after completing my day's labor." However, her record was bested by another entrant in the mileage contest: Miss Vera H. Merrian of the Red Cross Hut in the Hospital Area, who announced "that she danced eight and three-quarters miles [one] week." Other women staffed the Y's clubs in Beaune: one for officers and one for enlisted men.[36]

TRADITIONAL ACADEMIC COLLEGES

NOTES

1. See unsigned Bulletin No. 1, College of Business, A.E.F. University, March 10, 1919, in folder "College of Medicine, A.E.F.U.—Faculty Minutes, Orders, Memos," Box 1929, Entry 419. The College of Business not only based its organization and program on traditional business programs at American colleges and universities, but also on the lengthy, detailed, Bulletin No. 20, GHQ, AEF, March 10, 1919, *United States Army in the World War*, vol. 17, pp. 213-20, which was concerned solely with business education in the AEF's schools.

2. *A.E.F University News*, vol. 1, no. 5, May 22, 1919.

3. *Stars and Stripes*, May 16, 1919; *A.E.F University News*, vol. 1, no. 4, May 15, 1919. Even the clubs were provided for by Bulletin No. 20, cited above in note 1.

4. *A.E.F University News*, vol. 1, no. 4, May 15, 1919.

5. See undated, unsigned short sketch of the history of the Medical College; various documents on requirements for admission to study medicine or dentistry; and other reports, in unlabeled folder, Box 1927, Entry 419.

6. See, for example, memorandum from the adjutant general, GHQ, A.E.F., to president, American E.F. University, March 22, 1919, ordering Reeves to send eleven advanced students to Camp Knotty Ash, Liverpool, England, for further assignment to English universities for post-graduate level work in medicine, and four others to the University of Montpellier for duty as medical students, in folder 182, Box 1925A, Entry 419. Regarding the end of the Pre-Medical College, see memorandum No. 3, Headquarters, American E.F. University, March 8, 1919, in Box 1944, Entry 415.

7. Good photographs of the laboratory, classroom, and office facilities at the Medical College are in folder "College of Medicine A.E.F.U. Photographs," Box 1929, Entry 419. These reveal the essentially crude but adequate facilities provided. The classrooms contained straight-back, folding chairs, with no arms or desks. These were no doubt uncomfortable, and made the taking of notes difficult. The labs also were crude wooden benches but seem adequately provided with the usual lab equipment. There is also a photograph of the Medical College faculty, all officers, unlike some of the faculties that included enlisted personnel as instructors.

8. See several sheets of grades for various dental courses in folder "College of Medicine, A.E.F.U. Correspondence 26-50," Box 1930, Entry 419. These varied considerably. For example, in Captain Joshua H. Gaskill's class in fractures and splints, the twenty members of the class averaged a D as a final grade. The twenty-seven students in Lieutenant Morris Bernard S. Fleischer's course in oral hygiene, however, averaged an A.

9. Memorandum from the adjutant general, GHQ, A.E.F., to president, A.E.F. University, March 31, 1919, ordering Reeves to transfer thirteen students to the University at Edinburgh, via Camp Knotty Ash, Liverpool, England, for postgraduate work in veterinary surgery. These students were advised that the fee for their course was fifteen guineas (315 shillings), and that no student was to be sent unless he was fully prepared to meet the expenses out of his own pocket. See in folder "College of Medicine, A.E.F.U. Correspondence 151-175," Box 1930, Entry 419.

10. *A.E.F University News*, vol. 1, no. 5, May 22, 1919.

11. Ibid., vol. 1, nos. 4 and 5, May 15 and 22, 1919. See minutes of the

meetings of the faculty of the College of Fine and Applied Arts, for March 14, April 5, and April 30, 1919 in folder "College of Fine Arts and Applied Arts, etc. Faculty Minutes, Orders and Memos," Box 1924, Entry 408.

12. See undated, unsigned, "Brief Statement of the Work Done by the College of Engineering," in folder "College of Engineering. Its Work and Organization," Box 1934, Entry 419, and term reports for the College of Engineering in Box 1913, Entry 419. These reveal that on average 1/3 to1/2 of those enrolled in the courses departed the university for home before the end of the quarter. However, in a class in topographical surveying, all fifteen of those enrolled completed the course with two earning A's, nine B's, and four C's.

13. See memorandum from Louis E. Reber, director, College of Engineering, to the commanding officer, A.E.F. University, March 3, 1919, in which he requested the flying in of several specified aircraft from the Air Service depot at Colombey les Belles. Reber specifically stated that "we have no intention of using these machines for flying, but wish to give instruction in the design, rigging, and assemblage of the whole machine." This is in folder "Aeroplanes," Box 1908, Entry 408.

14. A.E.F University News, vol. 1, no. 6, May 30, 1919.

15. Ibid.

16. Memorandum containing report on the College of Agriculture, from Harry Hayward, director, to educational director, American E.F. University, Beaune, May 29, 1919, in folder "College of Agriculture, A.E.F.U.–Its Work and Organization," Box 1939, Entry 419, and several photographs of various activities at the College of Agriculture in folder "College of Agriculture–Photographs," ibid.; term reports for the College of Agriculture in Box 1913, Entry 419; and A.E.F University News, vol. 1, no. 4, May 15, 1919.

17. A.E.F University News, vol. 1, no. 4, May 15, 1919.

18. See memoranda, from head of the Department of Chemistry to the director, College of Science, March 16, 1919; and from Headquarters, American E.F. University, to chief, Chemical Warfare Service at Tours, France, March 7, 1919, in folder 179, Box 1925A, Entry 419; report on the history of the College of Science, in folder "College of Science, A.E.F.U.–Its Work and Organization," Box 1927, Entry 419. See also collection of drawings of bones and bone structures made by students of the college, perhaps for some pre-med course administered by the College of Science, in Box 1927, Entry 419; and A.E.F University News, vol. 1, no. 6, May 30, 1919.

19. See Roster of Administrative Staff and Faculty, College of Journalism, in folder "B," Box 1932, Entry 419.

20. There is much material on the college and the courses taught, course outlines, examination questions, and exercise assignments in folders "B" and "College of Journalism–A.E.F.U. Its Work and Organization," Box 1932, Entry 419. Fogg was one of the directors greatly concerned with his students' getting credit in American institutions for work done at Beaune. He therefore stressed that "the grading of students in this College is a matter demanding our *careful attention*." The instructors must set standards and adhere to them, "avoiding notable variation." In addition, care was to be taken during examinations, though "without ostentatious policing," to "guard the integrity of this work." See memoranda from Fogg, to instructors to the College of Journalism, regarding grades and the conduct of mid-term examinations, April 1 and 23, 1919, in folder "College of Journalism–A.E.F.U.–Faculty Minutes, Orders, Memos," Box 1932, Entry 419.

TRADITIONAL ACADEMIC COLLEGES

21. See memorandum, from the president, American E.F. University, to director, College of Journalism, April 5, 1919, for details as to how the páper was to be produced and what it was to include, in folder "College of Journalism—A.E.F.U.—Correspondence relative to the 'News,'" Box 1931, Entry 419.

22. *Stars and Stripes*, May 16, 1919. See article on the College of Journalism in *A.E.F University News*, vol. 1, no. 6, May 30, 1919. Reeves had hoped that the paper might be published locally in Beaune, but the printer there proved difficult, and negotiations were entered into with the Dijon firm. See several documents in folders "The A.E.F.U. 'News,'" and "College of Journalism—A.E.F.U.—Correspondence relative to the 'News,'" Box 1931, Entry 419.

23. For details of the College of Journalism, see memorandum from Fogg to the educational director of the university, May 31, 1919, containing a report on the activities of the College of Journalism, in folder "Memos, Bulletins, Special Orders, etc.," Box 1932, Entry 419. There are several photographs of the faculty and staff of the college, as well as of buildings and classrooms, in Entry 419, Box 1932, Folder "College of Journalism—A.E.F.U. photographs."

24. *A.E.F University News*, vol. 1, no. 6, May 30, 1919. See term reports for the College of Education in Box 1913, Entry 419. These reports reveal how fluid the class enrollments were as men returned to their units for the journey home. For example in one class of seven students, Special Methods in Agriculture, only one grade—an A—was recorded. All other members left for the United States before the class was completed. In other classes, about 1/4 to 1/3 was the average of those leaving the university early.

25. *A.E.F University News*, vol. 1, no. 4, May 15, 1919.

26. See memorandum containing a report of the College of Education from Joseph M. Gwinn, director, to Colonel Ira L. Reeves, May 31, 1919, in folder "College of Education AEFU Its Work and Organization," Box 1933, Entry 419; and *A.E.F University News*, vol. 1, no. 5, May 22, 1919.

27. See several photographs of faculty and students in the College of Letters, in folder "No. 4—College of Letters, A.E.F.U. Photography," Box 1930, Entry 419.

28. *A.E.F University News*, vol. 1, no. 4, May 15, 1919.

29. Ibid., vol. 1, no. 6, May 30, 1919. See also College of Letters Term Reports in Box 1913, Entry 419. These reveal the large enrollments in French. See large unlabeled folder ibid., containing the grades in French as turned in by the French soldier-instructors.

30. Much of the following is derived from an undated report entitled "History Of The College Of Music," by the college's director, Franklin Whitman Robinson, in folder "College of Music—Its Work and Organization," Box 1933, Entry 419.

31. See, for example, a concert given on Sunday afternoon, May 4, 1919, in the Place Carnot, Beaune, where citizens of the town could hear the "March of Anzacs," by Lithgow; "Firefly" by Friml; "Beautiful Ohio," by Mary Earl; and many miscellaneous American melodies, as well as selections from "Remick's Hits." See details in notice in folder "College of Medicine, A.E.F.U., Faculty Minutes, Orders, Memos," Box 1929, Entry 419.

32. See Report of Classwork in History of Music, from Walter Squire to Franklin W. Robinson, and his Report on Classwork in Theory and Harmony, June 4, 1919, in folder "College of Music. Its Work and Organization. Reports of Director and Assistant," Box 1933, Entry 419. These reports contain detailed outlines of the course contents.

33. See roster of the administrative staff and instructors of the College of Music, May 22, 1919, ibid.

34. Undated report on the "History Of The College of Music," by Franklin W. Robinson, ibid. See also class rolls and grades in the various classes in the College of Music in unlabeled folder, Box 1933, Entry 419, and College of Music Term Reports in Box 1913, Entry 419.

35. Two articles in *A.E.F University News*, vol. 1, no. 6, 1919.

36. Ibid., vol. 1, nos. 4 and 6, May 15 and 30, 1919.

CHAPTER 4:
The Maturing University

THE SWIFT MATURING of the university was striking. Reeves turned his attention increasingly to academic concerns and delegated the supervision of much of the day-to-day military control to the university inspector.[1] Certain military matters had to be addressed. The school was part of the AEF and participated in its concerns. Among these were several investigations that were initiated following criticisms of certain organizations, operations, and procedures that had surfaced during the months of war. One probe involved the welfare organizations, including the YMCA, the American Red Cross, the Knights of Columbus, the Salvation Army, and the Jewish Welfare Board, though the YMCA was the main focus of this inquiry. All American personnel who had any complaints against any of these organizations were instructed to report them to the office of the university inspector.[2]

The student-soldiers were constantly reminded that the college atmosphere, which perhaps tended to laxness, must not detract from concerns about military appearance, bearing and conduct. Some of the men were appearing outside of their quarters with blouses, overcoats, or raincoats unbuttoned, for instance, leading to a stern reminder that all must comply strictly with army uniform regulations. Likewise, in response to a concern expressed by General Pershing, Reeves reminded the men that "*shoes will be polished.*"[3] And as was the case elsewhere in the AEF, and among soldiers in the United States, there were numerous instances of the wearing of unauthorized military decorations, and war service and wound chevrons. Certain soldiers sported these, no doubt to dress up otherwise drab uniforms and to reap the benefit of the general enthusiasm and praise for the returning veterans. The soldier-students at Beaune were sternly ordered to avoid such abuses.[4]

Similarly, there was concern about the policing of buildings and grounds. The men seemed particularly prone to litter the campus with wrappers, containers, orange peelings, and the like. It became necessary to double the guards in an attempt to stop the practice. All guards were instructed to arrest any persons littering, regardless of their rank, and to require each offender to devote one hour's time to general policing of the area. Reeves insisted that it "be thoroughly understood by all that the grounds will at all times be kept a model of orderliness and cleanliness."[5] To assist in this campaign, a University Sanitary Police Squad was established under the supervision of the University Sanitary Inspector. Members of the unit, identified by a brassard on their left arm bearing the letters "S.P.," had the authority to demand of any person committing "a sanitary offence," that it be remedied immediately. Any who refused could be arrested, as could anyone who committed a "serious sanitary offence," whether he corrected it or not.[6]

Other disciplinary problems followed a more-or-less normal pattern, with the majority of the miscreants being charged with violating Article of War Number 96, failure to obey orders; Number 93, stealing; and Number 61, being absent without leave.[7] Among other things, a number of enlisted men and members of the YMCA had reportedly been riding on the tops of cars and on the engine of the narrow-gauge railroad in Beaune, a dangerous practice that violated orders and caused difficulties with the French. These actions had to be halted.[8]

As might be expected, there were clashes and misunderstandings between the Americans and the French. In one instance, citizens of Beaune living in one neighborhood decided to publish their concerns in the local paper, the *Journal de Beaune*. They complained of the "continual coming and going of Americans," and their perpetual oaths and loud laughter, which were highly offensive. They asked that the American authorities pay heed to their complaints, which was done: the street was placed off limits.[9]

Rather more entertaining on campus was the Friday evening "borrowing" of a guinea pig from a medical laboratory by unknown persons for some unknown purpose. Unfortunately, the animal had been inoculated with diphtheria germs for an experiment and it was urgent to locate it as soon as possible.[10] The matter was serious enough for university headquarters to issue a memo, No. 52, demanding the pig's immediate return to the medical laboratory. As a *Stars and Stripes* reporter observed, "A cross-examination of all mess sergeants has resulted in a universal plea of 'not guilty.' Corn Bill and goldfish, they say, are Corn Bill and goldfish, and you can't make a guinea pig taste like either."[11] And a writer for the *A.E.F University News* was certain that Parker Butler, author of a popular book, *Pigs Is Pigs*, would have made much of the incident had he been present at Beaune. But, in response to the official memo, the "guinea-cochon" was returned hurriedly by a Y man, "who modestly withheld his name from publication"; in fact, "when some of the medical staff went out to interview the individual regarding where he obtained the animal, said person gave an excellent exhibition of the Huns in the Argonne. All that was identified was a red triangle on his coat sleeve." In addition, another pig, not originally known to be missing, was returned by two YMCA girls "who, without knowing its true nature had adopted it as a mascot."[12]

A disturbance at the University Theater on the evening of Saturday, May 3, proved more serious. The enlisted men in the audience had been asked to move toward the rear, leaving the front rows open for about a thousand officers and ladies of the post who were expected to arrive for the evening's program. The men did so, but then responded by whistling and jeering the officers and Educational Corps personnel as they walked down the aisle to take their seats at the front. The officer in charge, Captain M. C. Carroll, canceled the evening's performance. The men dispersed, but once outside began making disparaging remarks against officers. They were called to attention and the facts of the incident were ascertained. Later, four men were formally charged with misconduct.[13]

There were other instances of disrespect to officers. A perennial problem in the AEF, this led in one instance to the confinement to barracks of a company of the 10th Provisional Regiment, though this punishment was later found to be unwarranted and was rescinded by higher authority.[14] But the feelings of the enlisted men toward officers were intensified by officious attitudes and by questionable conduct on the part of some officers. At Beaune, there were several instances of officers interfering with working parties. On one occasion, an officer put men who were working on a concrete detail on report for being out of uniform. In fact, the men were appropriately dressed in fatigue clothes. This case, and several others, led Major Richard Brooke, the district engineer at Beaune, to complain to Reeves, requesting that officers be instructed "not to interfere with men while at work." This was apparently done.[15]

Since the campus was a military installation, sooner or later a full-dress inspection by the commander-in-chief himself could be expected. Accordingly, on the morning of April 3, 1919, General Pershing's train arrived at Beaune's railway station, and was shunted to the university's siding. Some interruption of classroom work occurred, but the event was carried off successfully, to the school's general credit.[16] Present at the main address given by Pershing at Pershing Field were several French officials—Georges Crimanelli, *sous-préfet* of Beaune; Galotin Labrely, president of the Tribunal; and M. Roberjob, procurer of the French Republic—together with a total of 10,973 officers, enlisted men, nurses, and YMCA personnel of the Beaune post in formation.[17]

Three weeks later, Pershing returned for a second inspection. On this occasion, he was accompanied by his young son, Warren; the secretary of war, Newton D. Baker; and high-ranking officers from the General Staff at Chaumont. Pershing ordered that the university continue functioning on a normal basis, so that the secretary of war could observe the students, professors, administrative staff, and troops in the performance of their daily routine. After luncheon, the inspecting party visited the clubs in Beaune and then proceeded to Allerey, to the Army Farm School, to visit classes in session there. Reeves had specified that when Pershing's party returned to the Beaune campus that there should be "a ball game in full blast. . .as the General will care to stay three or four innings of a base ballgame, or if thought best, repeat the foot ball game of today, and he will stay to see several periods." The inspection party departed shortly afterward by train, Pershing having been favorably impressed with the school, the generally well-turned-out appearance and discipline of its personnel, and its accomplishments as a unique institution of higher learning in a military setting.[18]

Athletics related to the general school experience, as well as serving military purposes. The school boasted two athletic fields named Carlisle and Pershing, which were in frequent use. The university's athletic program was headed by Major William Howard, the athletic officer, and was planned "so that all officers and men in the University will get enough out-of-door exercise to keep them in good condition."[19] Each provisional regiment had an athletic officer assigned to it, with officers in each company also designated to supervise the development of athletics in each company. Inter-company contests were stipulated for indoor baseball, volleyball, track, basketball, and doughboy and all-point contests. At the inter-regimental level, indoor baseball, baseball, soccer, volleyball, track, basketball, and doughboy and all-point contests were held. The regimental teams were to be selected on the basis of inter-company games. The university was authorized to field baseball, track, and tennis teams, which would participate in AEF championship tournaments. In addition, local tournaments were also held in tennis, wrestling, and boxing. The Knights of Columbus sponsored and staged a series of boxing tournaments.[20]

As the academic term wore on, athletic activity accelerated. An all-university tennis tournament was held starting on May 23, after all organizations had had elimination contests to pick their best men. Three from each provisional regiment in both singles and doubles were allowed, while all other organizations could enter one singles and one doubles team for every three hundred men or fraction thereof attached to their respective units.[21]

The university also hosted a major track meet on May 10, in which soldier-students enrolled in French universities participated.[22] This meet was a preliminary for the large AEF meet scheduled for Paris on May 21, where men were to compete for places in the Inter-Allied Games scheduled for June. The university squad had honed its skills at a track meet held earlier on May 5 with entries from the Farm School at Allerey, with the Beaune group making a good showing. Indeed, on several occasions, the athletic encounters between the two schools produced some lively competition.[23]

The popular Sunday outings were continued as the term evolved. However, the Committee on Excursions decided that the men might wish henceforth to take them individually or in their own groups rather than in a single large body, as was the case before April 7. Therefore, a series of seven excursions was planned and described. The committee recom-

8. Gas Engine Class.   111-SC-153648

mended that maps of the area—*cartes d'état-major*—which were available in Beaune's bookstores, as well as a soon-to-be-published booklet on Beaune and its environs, which would be available in English, be obtained, to prepare the hikers. As before, the men were allowed to draw their luncheon rations; also, as always, they were on their honor to return to camp by 5:00 P.M.[24]

Another sign of the maturing school was its growing concern about publicity. Though the university had earlier issued several press releases to such papers as the Paris edition of the *New York Herald*, the *Daily Mail*, the *Chicago Tribune*, and the *Stars and Stripes*, and had appointed a publicity officer, various reports of misleading and malicious accounts about the university in newspapers in the United States and elsewhere began to surface.[25] For instance, in one Tennessee newspaper, an article alleged that the Beaune school "was merely a camoflouged [sic] penal institution of some sort or a depot for troops who had contracted certain unpleasant diseases."[26] These statements plainly violated the Espionage Act. All persons who had knowledge of such attacks were ordered to report them to the executive officer.[27]

To assist the university further in developing favorable publicity, as well as to provide a medium for instructing journalism students, the *A.E.F University News* was launched as a weekly student newspaper, under the direction of the director of the College of Journalism, Professor Miller Moore Fogg.[28] Correspondents were appointed by various units to turn in to the paper items of interest to the university community. Until a printing press could be shipped from Paris, the *A.E.F. University News* had a substitute: the *Bulletin of The A.E.F University News*, a mimeographed paper of eight pages. The first three numbers of the *News* appeared in the form of the *Bulletin*. When the press failed to arrive in time from Paris, the last four issues of the publication were published in Dijon. These were four-page papers, printed in six columns, and were packed with much information about the functioning of the university. Each press run was five thousand issues. The first issue from Dijon was No. 4, of May 15, the last being No. 7, dated June 6, 1919. In addition, because the school was not getting sufficient exposure at home—at least not of the favorable kind—university personnel were requested to send copies of the paper home.[29]

A major development regarding the university came in April when the YMCA relinquished its share of control and responsibility for all educational work in the AEF, including that at Beaune. The Y had initially established an educational system for the AEF "at a period when the

Army itself had to dedicate its entire personnel and resources to crushing Prussianism." Subsequently, operating mainly under General Orders 9 and 30, GHQ, AEF, the Y had proceeded to recruit and bring to France several hundred of the nation's ablest public school superintendents, school and college teachers, and professors. This staff was placed under the direction and control of the YMCA Army Educational Commission, which was charged with assisting the 5th Section of the AEF's General Staff, at Headquarters, Chaumont, in the development of the army division and post schools in France and Germany, and significantly participating in the development of the American E.F. University in Beaune and its branch work at Bellevue and Allerey. Then, in the era of peace, the AEF had established an educational system as an integral part of its structure and activities. It now seemed appropriate for the army to assume complete responsibility for the Army Educational Commission and its staff, and, in addition, to purchase the textbooks then being used throughout the AEF.[30] The army thereupon drew up a contract, which the Y's headquarters in Paris hoped that all of their people would "promptly and enthusiastically" accept, as most of them did.[31] Individual contracts extended from a minimum of two months to a maximum of six. The army agreements were similar to the Y contracts originally signed, and provided for a home allowance, a living allowance in France, a monthly allowance for uniform renewal of ten dollars (fifty francs), and sixteen dollars a month for insurance. In addition, when assigned to duty away from their regular post, as was the case with certain lecturers, they received four dollars per diem.[32]

The Y's Army Educational Commission and its personnel thereby gave way to the Educational Corps, AEF, which became official on April 16, 1919. The corps was headed by an Educational Corps Commission, made up of the same three men who headed the Y's Army Educational Commission, and who would remain at Beaune in their new capacity. This commission was charged with the development of educational policies in the AEF, standardization of educational methods, development of courses and course material, and general supervision over personnel of the Educational Corps, and assignment of its members. These people were to be ordered to specific army, corps, division, Services of Supply sections, or to the American E.F. University for general duty as instructors and as advisors to organizational commanders and their designated school officers. The army provided them with office space, issued their travel orders, and paid their travel expenses at the rate of four dollars per diem.[33]

SOLDIER-SCHOLARS

There remained the provision of the uniform and the adoption of appropriate insignia. General Orders No. 68, GHQ, AEF, Chaumont, April 19, 1919, stipulated that the uniform for members of the Educational Corps was to be that of a dismounted commissioned officer. In addition, the members were authorized to wear the Sam Browne belt, and the cap insignia worn by a commissioned officer, though without insignia of rank. The new unit's branch insignia consisted of a 2 7/8 by 3 1/8-inch khaki shield with an insert consisting of a winged torch device and the designation "Educational Corps" emblazoned on it. The colors dark blue, gold, and red were employed. This insignia was worn on the right sleeve with top of the shield one-half inch below the seam of the shoulder. They wore the same insignia on the right side of the cap. A bronze ornament for the collar would be employed as soon as available. Corps members were also to have the same courtesies extended to them as those normally accorded commissioned officers.[34]

MATURING UNIVERSITY

NOTES

1. General Orders No. 38, Headquarters, American E.F. University, May 9, 1919, Entry 412.

2. General Orders No. 39, Headquarters, American E.F. University, May 15, 1919, Entry 412. The YMCA had been accused by American soldiers of overcharging troops for canteen items. Since the public had contributed generously to the Y's efforts, to the tune of almost 162 million dollars during several fund-raising campaigns, there was a public outcry as well as demands for an investigation. The War Department and Congress responded, and hearings and investigations were conducted. The Y learned that it had overcharged on certain freight costs, but it was generally exonerated, as were the other organizations under investigation. However, the Y's reputation suffered as a result of the adverse publicity. See Young Men's Christian Association, National War Work Council, *Summary of World War Work of the American Y.M.C.A.* (New York: International Committee of the YMCA, 1920), pp. 161-62; 205-6. Pershing's aide, Colonel George C. Marshall, was one of those in France who felt that the criticisms of the Y were unjustified. "With a keen sense of injustice, [he] persuaded General Pershing to come publicly to the defense of the Y and later he himself took every chance he got to set the record straight." See in Forrest C. Pogue, *George C. Marshall: Education of a General* (New York: Viking, 1963), p. 196.

3. Memoranda Nos. 50 and 54, Headquarters, American E. F. University, May 8 and 12, 1919, Entry 415. Pershing's preoccupation with shined shoes was proverbial. The number of official memoranda and orders from GHQ, Chaumont, containing references to this subject attests to this, as do numerous allusions to it throughout his career. Indeed, as one observer noted, "he had an awesome devotion to shining shoes, a devotion bordering on fetish." See in Frank E. Vandiver, *Black Jack: The Life and Times of John J. Pershing*, 2 vols. (College Station: Texas A&M University Press, 1977), 1: 111.

4. General Orders No. 43, Headquarters, American E.F. University, June 2, 1919, Entry 412.

5. Memorandum No. 34, Headquarters, American E.F. University, April 15, 1919, Entry 415.

6. Memorandum No. 31, Headquarters, American E.F. University, April 12, 1919, Entry 415.

7. See notebook pages "Index Charge Sheets," for the Farm School at Allerey, in bound notebook, "Sub-Post Hqrs. A.E.F.U. at Allerey," in Box 1964A, Entry 420. This lists a total of forty-seven charges for the school term against personnel at Allerey. Some of the men charged were determined to be not guilty. Many were sentenced to a fine of $7.50 each month for two months; some were reduced in rank; some spent short terms in confinement, typically for ten days or one month, some at hard labor. Apparently no serious charges were brought against men enrolled at the Farm School. See other details of actions taken against Beaune and Allerey personnel in folder "Charges against enlisted men," Box 1919, Entry 408.

8. Memorandum No. 53, Headquarters, American E.F. University, May 11, 1919, Entry 415.

9. See the *Journal de Beaune*, April 23, 2929, and relevant documents in folder "Investigation by I.G. and Other Officers," Box 1922, Entry 408.

10. Memorandum No. 52, Headquarters, American E.F. University, May 10, 1919, Entry 415.

11. *Stars and Stripes*, May 23, 1919.

12. Ibid., May 30, 1919; *A.E.F University News*, vol. 1, nos. 4 and 5, May 15 and 22, 1919.

13. See relevant documents in folder "Military Reports, HQ, 1919, American E.F. University," Box 1910, Entry 420.

14. The relevant documents are in folder "Investigation by I.G. and Other Officers," Box 1922, Entry 408. This case led Dallam to issue a memorandum ordering commanding officers to pay strict attention to administering disciplinary action in compliance with the Articles of War, and the Manual for Courts-Martial. See Memorandum No. 48, Headquarters, American E.F. University, May 5, 1919, Entry 415.

15. Brooke also suggested that all officers be instructed to read paragraph 382, section 3, Army Regulations. See Memorandum, Major Richard Brooke to Colonel Reeves, in folder "Conflict of Officers Authority," Box 1903, Entry 408.

16. Memorandum No. 23, Headquarters, American E.F. University, April 2, 1919, Entry 415, for details of entire program to be followed. See also *Stars and Stripes*, April 11, 1919.

17. See Report of Officers, Enlisted Men, Nurses, and YMCA Present At General Pershing Review, April 3rd, 1919, in folder "Ceremonies," Box 1923, Entry 408, and *Stars and Stripes*, April 11, 1919.

18. Undated memorandum, from Colonel Reeves to the Executive Officer, A.E.F. University, Beaune, in folder "Ceremonies," Box 1923, Entry 408. In addition, members of the United States House of Representatives' Committee on Military Affairs inspected the university on May 3. *A.E.F University News*, vol. 1, no. 4, May 15, 1919.

19. *Stars and Stripes*, April 18, 1919. Howard had previously worked for six years as an athletic director in New York City.

20. Bulletins Nos. 50 and 68, Headquarters, American E.F. University, April 2 and 18, 1919, Entry 414. See also Bulletins Nos. 113 and 119, Headquarters, American E.F. University, May 27 and June 3, 1919, Entry 414, for details relating to the Knights of Columbus boxing bouts.

21. Bulletins Nos. 68 and 89, Headquarters, American E.F. University, April 18 and May 15, 1919, Entry 414.

22. Bulletin No. 78, Headquarters, American E.F. University, May 2, 1919, Entry 414.

23. Articles on the May 5 and 10 meets are in *Bulletin of The A.E.F University News*, vol. 1, no. 3, May 9, 1919.

24. Bulletin No. 61, Headquarters, American E.F. University, April 11, 1919, Entry 414.

25. Memorandum No. 5, Headquarters, American E.F. University, March 10, 1919, Entry 415. See article by F. Wythe Williams, in the *Daily Mail*, Paris, March 3, 1919, entitled "U.S. Army Magic University," which was a quite favorable account of the school's establishment and its ambitions, and others in folder "Publicity and Public Press," Box 1901, Entry 408, and Press Release, March 10, 1919, in folder, "Newspaper Clippings," Box 1937, Entry 419.

26. See memorandum, Captain Benjamin Brooks, to Colonel Charles W. Exton, Beaune, May 6, 1919, in folder "Publicity and Public Press," Box 1901, Entry 408.

27. Memorandum No. 49, Headquarters, American E.F. University, May 6, 1919, Entry 415.

28. Fogg had been active on the home front before sailing for France, having served for some months as the Nebraska state chairman for the "Four Minute Men," speakers who delivered patriotic speeches in movie theaters and elsewhere seeking to rally public support for the war effort. For the Four Minute Men, see Alfred E. Cornebise,

MATURING UNIVERSITY

War As Advertised: The Four Minute Men and America's Crusade, 1917-1918 (Philadelphia: American Philosophical Society, 1984).

29. Bulletin No. 102, Headquarters, American E.F. University, May 21, 1919, Entry 414.

30. Letter, from E. C. Carter, chief secretary, YMCA in the AEF, Paris, to General John J. Pershing, commander-in-chief, AEF, March 14, 1919, in folder "A.E.F. University. Historical Data," Box 1956, Entry 409; Letter from General Pershing to E. C. Carter, GHQ, AEF, March 25, 1919, ibid.

31. Letter to all Y Educational Staff, from E. C. Carter, chief secretary, YMCA, AEF, Paris, March 27, 1919, in folder "Y.M.C.A.—Army Educational Commission," Box 1956, Entry 409.

32. Letter from Frank E. Spaulding, commissioner on the Y's Educational Commission, to field representatives, Army Educational Commission, April 12, 1919, in folder "Personnel, A.E.F. University—Instructors," Box 1956, Entry 409.

33. General Orders No. 62, GHQ, AEF, April 8, 1919, *United States Army in the World War*, vol. 16, pp. 725-29. This order revoked the relevant paragraphs of General Orders Nos. 9 and 30. See in folder "Educational Corps, A.E.F.," Box 1956, Entry 409. The transfer of the civilians to the Educational Corps, AEF, was effected by Special Orders No. 105, GHQ, AEF, April 15, 1919. This contains a complete list of those involved. See copy in folder "General Correspondence of Ira L. Reeves," Box 1956, Entry 409.

34. General Orders No. 68, GHQ, AEF, April 19, 1919, *United States Army in the World War*, vol. 16, pp. 734-36; Memorandum No. 29, Headquarters, American E.F. University, April 10, 1919, Entry 415; Bulletin No. 66, Headquarters, American E.F. University, April 17, 1919, Entry 414.

CHAPTER 5:
The Farm School

When Colonel Reeves set out on February 7 to inspect the former hospital centers at Allerey and Beaune, he decided that the site at Allerey (Saône-et-Loire), located about ten miles southeast of Beaune, was unsuitable for the establishment of the main campus of the proposed AEF University, though it might do for an agricultural school.[1] Nevertheless, he was not pleased with it. He noted that "the impression one gains when arriving at the site is anything but favorable." The general situation was poor. The roads, sidewalks, and the drainage system were inadequate and unsightly, and the ground was quite muddy. However, the buildings of the former hospital complex, one of the largest base hospitals in the AEF, could easily accommodate about eighteen thousand soldiers. They were well constructed, cooking facilities were ample, and with a large force of laborers and a sufficient number of engineers, the site could be greatly improved, he concluded, though the general environment and the low-lying location would, of course, remain. Despite his reservations, GHQ decided that it would be adequate for the campus of the Army Farm School.[2]

Such a school would certainly be welcomed in the army, as there was little doubt of the interest in agriculture among the doughboys. Indeed, agriculture was considered one of the most important subjects in the army's educational programs. One of the people responsible for bringing some of this interest into focus was Kenyon Leech Butterfield, of the Y's Army Educational Commission. He was in charge of vocational educational work in the AEF, which included his first love, agriculture. Before his arrival in France, he was president of the Massachusetts Agricultural College at Amherst, and had long experience in farming and the study of agriculture. Born in Michigan on a dairy farm, he entered the Michigan State College of Agriculture, graduating in 1891. He pursued graduate work at Amherst, which in 1899 awarded him the honorary degree of doctor of laws. He received an M.A. from the University of Michigan in 1901. For a time, he served as superintendent of farm institutes and as editor of a Grange publication. In 1903, he became president and professor of political economy at Rhode Island

THE FARM SCHOOL

Agricultural College. Three years later, he became president of the Massachusetts Agricultural College, coming to France on a leave of absence from that school. One of his interests was the farmer and his family, which, as he was fond of remarking, are of more consequence than the farm itself.[3] It was Butterfield who initiated many of the programs to instruct the men of the AEF in agrarian subjects.

In compliance with Bulletin No. 9, GHQ, which stipulated that farmers' clubs, institutes, and short courses in agriculture would be provided throughout the AEF, by mid-January, in and around Bordeaux, three-day farmers' institutes were conducted in twelve camps in the area. Subjects addressed included soils, livestock, farm management, and how to secure and finance a farm. In addition, a night course was set up for the benefit of labor troops. It featured lectures and movies furnished by the Department of Agriculture, emphasizing the obtaining and financing of farms in the United States.[4] The institutes spread rapidly, and by early May, some 220 had been held, reaching over one hundred thousand men.[5]

The club movement also developed in an impressive fashion throughout the AEF. These organizations provided a convenient way to focus the agricultural interest of the farmer-soldiers and give them an opportunity for continued study and discussion of agricultural subjects, especially at isolated posts lacking other educational opportunities. The Army Educational Commission promoted these developments, providing information and encouragement. The clubs were often formed following short courses, institutes, or lectures, or simply through interest stirred up by school officers or by the men themselves. Numerous successful organizations were soon in operation. For example, in the 91st Division, awaiting shipment home at its camp at Le Mans, one regiment recorded that more than thirteen hundred men were enrolled in farmers' clubs, and were determined to continue their activities on shipboard, until "the shores of the U.S. were reached." Indeed, the base and embarkation areas, such as Le Mans, Brest, Bordeaux, and St. Nazaire, were especially active centers for farmers' organizations. These clubs, usually composed of ten or more men, held regular meetings featuring discussions and debates, brought in guest speakers and lecturers, and arranged trips to observe French agricultural methods in the countryside surrounding military establishments. The subjects considered included marketing, rural credit, farm community life, and the place of the farmer in national and world affairs.[6]

The various post and divisional schools also normally offered agricultural instruction in several fundamental courses, lasting from six to twelve weeks. An exceptional divisional school in this regard was that of the 89th Division, which offered twelve regular courses in agriculture.[7]

A further enlargement of agricultural instruction came when GHQ issued General Orders No. 28, on February 11, 1919. This instituted elaborate plans for the provision of vocational training throughout the army. The Army of Occupation in Germany, for example, was pleased to note, along these lines, plans for creating a training center in the region of Prun, Neurburg, and Saarburg, where German agricultural schools were already being utilized by the 89th and 90th divisions. Men from the agricultural states of Texas, Oklahoma, Kansas, and Missouri formed the larger part of these divisions.[8]

Other opportunities that were available to soldiers of the AEF were provided by several French agricultural colleges that accepted a limited number of advanced students. Also, one of the colleges at Beaune was the College of Agriculture, headed by Butterfield. However, the college was intended to accommodate students who wished to study agriculture more formally and at an advanced level. What was needed was a school to establish a more general course in agriculture that would serve the doughboy who had a limited knowledge of agriculture, and who desired to become involved in agriculture following his return to the United States. The Army Farm School at Allerey met these needs.[9]

Initiators of the new institution were Reeves, Butterfield, and Herbert J. Baker, who became the school's principal. Created as a sub-post of the University at Beaune, its first commanding officer was Major A. C. Judd of the 310th Infantry, who was relieved on April 12 by Major Andrews H. Patterson, late of the 323d Infantry. The commanding officer of the sub-station was on the staff of the superintendent at Beaune, his official title being "Executive officer, Sub-station American E.F. University, Allerey." His staff consisted of an adjutant, a personnel adjutant, a superintendent of buildings and grounds, and a sanitary inspector, with other staff assigned as necessary. Eventually an inspector general, a military director, an assistant executive officer, a registration officer, a details and transportation officer, an entertainment officer, a mess officer, and a fire marshall were added.[10]

The number of students that the new school could accommodate was set at 2,993, resulting in an allotment system whereby men from each of the military units still in Europe could be sent by quota to Allerey. In this system, each division averaged just over one hundred men, though the Services of Supply received an allotment of 337.[11]

THE FARM SCHOOL

In addition to the students, other soldiers, many with farm experience, were detailed to Allerey to help operate the new facility. Many of these were from the 323rd Infantry Regiment of the 81st Division.[12]

As was the case on the main campus at Beaune, the educational effort took precedence over the military aspect, and in short order the instructional apparatus was in place.[13] The school's principal was Herbert J. Baker, of the Army Educational Corps. He held a B.S. degree from the Massachusetts Agricultural College, and had been the director of extension work for the state of Connecticut. The vice principal was initially Captain Russel T. Gheen, a field artillery officer, but he was soon replaced by Captain Roscoe R. Snapp, of the 112th Field Artillery. Holding a B.S. degree from the University of Illinois, in civilian life Snapp was an instructor at the Illinois College of Agriculture. The registrar, Captain Rubey J. Hamilton of the 123rd Field Artillery, was a teacher in civilian life. His B.S. degree was from the University of Illinois. He also taught in the agronomy department.

Under these three, and a man at Beaune, were four administrative divisions. One was the Class, Laboratory and Study Division, headed by Thomas E. Elder, a member of the Army Educational Corps who also acted as the major student advisor. His B.S. degree was from Cornell and he was formerly an agricultural director at a prep school. The second was the Division of Practice Work. Its superintendent was 2nd Lieutenant George N. Danforth, yet another field artillery officer, who was from the 104th Field Artillery. His B.S. degree was from the Massachusetts Agricultural College. In civil life he was a school teacher. The third administrative division was that of Recreation and Organized Athletics, which was under the supervision of Major William Howard of the Beaune campus. The fourth and final division was the School Farm. The farm superintendent was 1st Lieutenant Marshall W. Sergeant, of the 320th Infantry. A farm manager in civilian life, he held a B.S. from Wisconsin. Lieutenant Sergeant had three assistant farm superintendents under him, one for farm animals, another for horticultural crops, and a third for field crops.[14]

Elder headed the six academic departments of the Farm School, which were agronomy, animal husbandry, horticulture, agricultural economics and sociology, agricultural engineering, and English. The formal courses of study were: agronomy, animal husbandry, horticulture, farm equipment, economics and sociology, and shop work. The English department concentrated its efforts on courses for illiterates and elementary-level students.

The Farm School's faculty and staff were well qualified for the level of instruction given and the goals that the school sought. The school eventually had ninety-two instructors, of whom thirty-one were commissioned officers, fifty-four were enlisted men, and seven were members of the Army Educational Corps. A consideration of the qualifications of those who headed each department indicates something of the level of professional competence at the institution. First Lieutenant Herbert E. Drew was in charge of the Department of Animal Husbandry. A member of the 807th Pioneer Infantry, he held a B.S. from the University of Wyoming and an M.S. from the Michigan Agricultural College. The Department of Horticulture was in the hands of 1st Lieutenant Leroy R. Frank, of the 324th Infantry. His B.S. was from Florida State University, and in civilian life he was a field manager for the Gulf Fertilizer Company. The head of the agronomy department was Captain Rubey J. Hamilton, who is discussed above. The agricultural engineering department's head was 2nd Lieutenant Richard C. Miller, a Transportation Corps officer. Holding the B.S. degree in engineering from Iowa State College, he was formerly an extension service agent at Ohio State University. First Lieutenant Thomas S. Bayne, of the 7th Infantry, head of the Department of Agricultural Economics and Sociology, held the B.S. degree from North Carolina State College. Before entering the army, he was principal of the Farm Life School in Lillington, North Carolina. Dr. Robert T. Kerlin, head of the Department of English, held the Ph.D. in English from Yale.

Not all of the faculty held college degrees; some made up the lack with practical knowledge of agricultural subjects and practices. Yet many were college-educated. For example, Private 1st Class Smith G. Beilby, from Headquarters Detachment of the 3rd Division, a high school teacher in civilian life, and an instructor in farm management at the Farm School, possessed his B.S. from Cornell. Second Lieutenant Clifford E. Dennis, of the 304th Infantry, an instructor of dramatics in the Farm School's English department, held a bachelor's degree from Harvard and an M.A. from Rutgers. Private L. P. Foster, of the Quartermaster Corps, a fruit grower in civilian life, with a B.S. from Ohio State College, was an instructor in horticulture. Farrier Clarence L. Goodnight, of the Veterinary Corps, though he possessed no degree, was a stockman in civilian life, and brought his practical knowledge to his classes in animal husbandry. Private 1st Class Thomas P. Metcalfe, of the 13th Field Artillery, held a B.S. from Texas A. & M. College and was in civilian life a livestock specialist in the animal husbandry division of the U.S. Department of Agriculture. He instructed in animal husbandry at the Farm School. Numerous other instructors were teachers in civilian life, while many others were field or extension agents in their home states, and whether officers, enlisted men, or members of the Army Educational Corps, shared common backgrounds in practical agricultural knowledge and experience, or academic training in the field. Predictably, a large number were from schools and colleges in the Midwest—Iowa, Illinois, Ohio, Minnesota, and Wisconsin especially—though the Massachusetts Agricultural College, no doubt reflecting Butterfield's presence and position, was well represented on the Farm School's faculty, as was Cornell.[15]

By mid-term, the Farm School's enrollment was 2,361 students, though initially more than 12,000 applications had been made for the school, an indication of the desirability of studying there. The student body was from every state in the Union, and included as well one student each from Alaska, Hawaii, Norway, and Sweden.

The school operated apart from the College of Agriculture at Beaune, though it was attached to it. It possessed 350 acres of land for school use, 250 of which were cropland. The school's farm machinery, implements, horses, and mules were supplied by the Quartermaster Corps, while cattle, sheep and poultry were contributed by the YMCA. The books were provided by the ALA as was the case, for the most part, at Beaune. The collection totaled approximately 6,000 volumes, with an average circulation of from 350 to 400 volumes a day.[16]

The aims of the school were to give a practical course in agriculture to those who desired to enter farming following their return home; to acquaint men of limited experience with various kinds of farming, thereby assisting them to choose one, and to identify an area of the United States to settle in; and to teach the advantages, problems, and opportunities of rural life, while promoting an appreciation of farming as a way of life.[17] The school had no special educational requirements for entrance, and any member of the AEF interested in a practical course in agriculture could apply. The students were also permitted, as at Beaune, to return to their organizations for embarkation when they departed. Those students completing the course of study in a satisfactory manner were awarded an American E.F. University Certificate to that effect. The program was so arranged that all students were required to take three or four subjects for a period of three weeks. Then, another group of subjects was taken until all the courses offered had been covered. The students attended lectures, did laboratory work, and were given practice work on the farm. Each student spent from three or four hours a day in the classroom or laboratory, two hours a day in practice work, an additional hour in supervised study, and two hours a day in recreation. Free discussion was allowed, both during the lecture period and at the end of the class. This discussion was "entered into heartily," and because all sections of the country were represented in the classes, bringing forth different opinions and experiences, it was "one of the best features of the class-room work." While some officials, felt that the course load was too heavy, the students did not object to it, and "the quality of the work done was most satisfactory."

It was also intended that a series of short courses of one month's duration would be offered to other soldiers who had not been able to enroll at Allerey. A faculty was recruited and organized, and a course of study prepared. Orders were sent out to fifteen hundred students to participate, but because of the acceleration in transporting troops to the United States the short courses were canceled.[18]

The students and staff were comfortably housed in wooden barracks formerly used as hospital wards, so it was not necessary to construct any new buildings. The classrooms and laboratories were also converted hospital units, and some of the existing structures were modified for use as livestock buildings, making it necessary to put in heavier floors and mangers. Of the farm's 250 acres of tillable land, about 150 acres were plowed and sowed in various crops, though the problems to be overcome were considerable. The farm superintendent complained that the arable land was at first mostly mud and that none of it was properly drained. Some of it had not been cultivated for seven years, and there was no fertilizer to put it into shape. In addition, there were several hundred tons of brush, logs, and stumps scattered over it. Nonetheless, with the help of a willing student force, the difficulties were overcome: a mile-long drainage ditch was cut, and much of the land was plowed and planted.[19] The fields, orchards, and vineyards in the vicinity were also used for demonstration purposes by the school. The livestock consisted of eighty-two horses and mules, of which thirty were used for farm work, the remainder for student practice. The original dairy herd consisted of eleven animals, to which were added fifty swine and over one hundred chickens. Livestock associations in various parts of Europe also lent purebred Percheron and French coach horses, Jersey and Guernsey dairy cattle, and various breeds of sheep, hogs and poultry. The farm was equipped with the most modern farm machinery, much of it selected from machines brought to France by the U.S. Garden Service, which had earlier planted crops and gardens to help feed the AEF.[20]

A large educational agricultural exhibit, illustrating various types of farming in different sections of the United States, prepared by the U.S. Department of Agriculture, was housed in an exhibition building. Designed to emphasize the practical needs and problems of the general farmer, it featured displays by the Weather Bureau, the U.S. Forest Service, and the bureaus of Animal Industry, Plant Industry, Chemistry, Soils, Entomology, and Public Roads, and included models of farm buildings and farm and household equipment and machinery.[21]

The organizing of learning at Allerey featured several innovations. These included clubs, which met two evenings a week to discuss topics of mutual interest. There were dairy, poultry, beef, and fruit clubs among others. Approximately 1,850 of the over 2,300 men enrolled were involved in these organizations. About thirty were established with an average membership of around sixty members each. The clubs were frequently named, the men in the three "hog clubs" naming them after the three principal breeds of hogs, and likewise the dairy clubs after the different types of dairy cattle. To one observer the men were commendably enthusiastic, their interest amounting "almost to a deep passion."[22] Nevertheless, the clubs had to be assiduously encouraged because they were soon meeting strong competition "in the outdoor sport programs of the army and the strenuous amusement campaigns under way." Therefore, it was asserted, a "special effort is needed to maintain them in the face of these things."[23]

In addition to the clubs, on two evenings a week the student body and faculty met in a thirty-minute assembly. Visiting speakers usually lectured during these periods, seeking to assist the men in choosing their life's work, in comprehending more fully the "meaning of citizenship," or in obtaining a better understanding of America's role in the modern world.

Another development that enhanced the instructional program was tours to nearby places of agricultural interest. Usually set for Sunday afternoons, these excursions, sometimes involving over 350 students, concentrated in a particular area or subject. The students visited notable gardens, forests (to study forest management), livestock production centers, dairies, flood control systems, and field crops. The men were accompanied by interpreters and lecturers. The participants were provided with mimeographed summaries and itineraries, and were encouraged to take cameras, though official photographs, of which prints were available to tour members, were also taken.[24]

The tours also afforded the students otherwise unavailable opportunities to study French ways of living. Harry Hayward, head of the Agricultural College at Beaune, stated that this aspect of the hikes had "been a most wholesome factor in helping both faculty and students to appreciate the people we have been living among while in France." Indeed, in his view, "the Sunday trips stand out as one of the important accomplishments of the School."[25]

The courses of study in agronomy, animal husbandry, horticulture, farm equipment, and shop work were more or less traditional. In economics and sociology, however, there were innovations. One of the aims of the school was to assist students in choosing where they might wish to live in the United States. Therefore, the course in agricultural geography emphasized the possibilities and difficulties of farming carried on in different sections of the nation. Attention was given to the lands still undeveloped, and to how these might be turned into profitable farms. The United States and its agricultural relations with the emerging world were also emphasized. Classes in rural sociology were formed into a Rural Community Club, which conducted the entire course of study in this format. Instructors and speakers addressed the club on such topics as agricultural colleges, the use of the Department of Agriculture's extension service, the function of the Grange in rural America, the role of the rural church, rural education, and other related topics. Throughout, the club considered itself a genuine rural community dealing with typical rural problems.[26]

Forty minutes each weekday were devoted to military drill, except for Saturday, when a two-hour military inspection was held. Two hours per day were set aside for recreation, and physical directors and coaches were available to assist the men in their exercise programs. Talent shows were also held. Indeed, the men at Allerey were almost as well entertained as were the students at Beaune. Many of the performances that reached the main campus arrived at Allerey as a matter of course, as, for example, the "Front Line Revue," a vaudeville show staged by soldiers, all of whom had been wounded in action. They had planned their show while convalescing in the hospital. The major part of their performance was "A Burlesque On a Day In a Hospital Ward." After a three-day stand at the University Theater in Beaune, the group opened a four-day engagement at Allerey. Other vaudeville acts at Allerey were "The Red, White and Blue Minstrels," a 41st Division show touring the AEF theater

THE FARM SCHOOL

circuit, and another group, "Tricks and Tunes," both of which also performed at Beaune. The "Tricks and Tunes" cast consisted of two women and one man, fresh from the United States, who performed the latest song hits from home, such as "Ja-Da," and "Till We Meet Again." They also did tricks and novelty acts. In addition, the Allerey school mounted its own variety show, which performed at several of the area Y huts on the Beaune campus and elsewhere. Numerous dances were held in Allerey, including a permanent schedule at the Nurses' and Officers' Club four nights a week.

A full athletic program was also instituted. Lieut. H. F. Purcell, assistant athletic officer of the university, was in charge of Allerey's athletics. The most important sport there was baseball, which was accommodated on seven diamonds at the Farm School, featuring a twilight league of ten teams. Indoor baseball participants had the use of twelve diamonds. Volleyball players had six courts at their disposal, and those who were interested in basketball had five courts available for their use. A fully-equipped track for track and field events was in operation. Several track stars emerged, and were subsequently placed on the university team at Beaune for competition with other university and AEF teams. Five tennis courts provided places of action for members of the tennis team that represented Allerey in the tennis competition circuit. Soon there were regular Thursday evening athletic contests between Beaune and Allerey, which included baseball games, track contests, and boxing and wrestling matches.[27]

Some students at the Allerey school, interested in journalism, also desired to establish unit or school newspapers. The 11th Regiment established the mimeographed *11th Regiment Bulletin*, the first two-page issue of which appeared on March 21. However, it soon ran afoul of Colonel Reeves when in its ninth number it published the results of a straw vote, apparently on the upcoming elections in the United States. Reeves acted swiftly, reminding the paper's editor that "it is strictly prohibited for officers and soldiers to take part in political matters, and this act is in strict violation of orders." He did not wish to suppress the publication, "as it no doubt has helped the esprit de corps of the regiment," but this action had caused him "to lose confidence in those having the publication in hand." He wanted the people responsible to be identified, and the bulletin suppressed, and would allow no further issue to appear until every article had first been censored by an officer designated by Captain Wright, the commanding officer of the 11th Provisional Regiment. Furthermore, Wright was ordered to take energetic measures to recall all copies of the offending bulletin and to destroy them. In addition, no letters were to be sent to the United States containing copies of such bulletins; if found, "such copies will be destroyed at once in your presence." This official response seems to have ended the paper's short career.[28]

Another newspaper venture fared little better. In early May, several individuals decided to launch a mimeographed paper for the entire Farm School. Informally called "The War Baby," the thirteen-page paper contained much interesting material. However, its very length worked against it, and Colonel S. Field Dallam, the university's inspector general, acting as the school's president while Reeves was temporarily absent in Germany, noted that the paper was in violation of the university's Memorandum No. 14, of March 22, 1919. This demanded the saving of paper, and Dallam pointed out that "you readily see that 50% of the paper used will be sufficient for this publication." Captain Roscoe R. Snapp, the Farm School's vice-principal, asserted that the failure to follow orders was "due to carelessness of [the] Student Editor of [the] paper," and indicated that "the mistake was noticed immediately after the paper was printed," and that precautions had been taken so that the error would not be repeated.[29]

The end of the Farm School came rather suddenly. Reeves informed Major Patterson that it had been decided to close the installation on May 31st. He ordered that every effort be made to clean up the buildings, to dispose of the property and the animals, and to ship out all university personnel prior to that date. A separate area at Beaune was reserved for all Allerey personnel.[30] In addition, all engineer personnel were to be evacuated by June 5, and on the following day, the camp at Allerey was to be transferred to the French army.[31] However, if anything, agricultural instruction intensified for members of both the Farm School and the College of Agriculture at Beaune in the few days remaining of the term. The Farm School closed out its formal courses with a Farmers' Institute meant to round out the courses already taken by giving all men an opportunity to get a few lectures on all subjects in the course. Originally scheduled to be presented at Allerey, the institute was moved because of the transfer of all personnel to Beaune, where it was held on June 2 and 3.[32] Simultaneously, an agricultural fair was opened. The exhibits were apparently those of the Department of Agriculture's set-up earlier at Allerey. The Farmers' Institute was immediately followed by a four-day Inter-Allied Conference on World Agriculture, which included numerous agricultural experts from many parts of the world.[33]

THE FARM SCHOOL

Though the Farm School at Allerey existed for only a short time, it clearly accomplished its purposes in a commendable fashion. It had instructed its enthusiastic student body in many aspects of theoretical and practical agriculture, and it had helped the students better to understand the French. In addition, it had also helped "to fix in all those associated with [it]. . . higher ideals of citizenship." There was little doubt of the close involvement of the students. It was often remarked by members of the faculty that indeed they had "never seen a group of students so keenly interested in their studies as were the students in this School." Another factor was no doubt important: the students were from all parts of the country. Their varied experiences greatly enhanced the instruction, and indeed were a liberal education in themselves, which greatly assisted the school in reaching some of its major goals.[34]

If it were true that to hundreds of the men the AEF University at Beaune would be the only alma mater they would ever know, it was equally true that to hundreds of others that alma mater would be the Farm School at Allerey. The experiences that they had in their only encounter with an institution of higher learning would remain with them for a lifetime. Surely for them, if not for many others in the AEF, their experience in the American army would end on a decidedly positive note.

SOLDIER-SCHOLARS

NOTES

1. *The Catalogue*, part 1, p. 4; Memorandum to Dr. John Erskine, educational director, American E.F. University, and chairman, Army Educational Commission, from Colonel Ira L. Reeves, May 13, 1919, folder "A.E.F. University, Historical Data," Box 1956, Entry 409. For a description of Allerey at the time of the hospital's operation, see Albert M. Ettinger and A. Churchill Ettinger, *A Doughboy with the Fighting 69th. A Remembrance of World War I* (New York: Pocket Books, 1992), pp. 196-221, n. 2; 317-18, n. 10; pp. 319-20.

2. See his detailed report to the assistant chief of staff, G-5, February 9, 1919, in folder "Allerey Sub. Station," Box 1922, Entry 408. Reeves recommended that the current commanding officer of the hospital center at Allerey be ordered to retain all personnel, then numbering about five thousand, and all supplies then in place, until the needs of the university and farm school should be ascertained. This was apparently done. His report contained a listing of the units stationed there and the major supplies at the site. He also included a status report on the Red Cross and its activities and possible expansion of its services if required.

3. *A.E.F University News*, vol. 1, no. 6, May 30, 1919. Reflecting this interest, the College of Agriculture at Beaune created a Department of Rural Sociology and Economy.

4. *Stars and Stripes*, January 24, 1919. These institutes, according to Bulletin No. 9, GHQ, issued later, were to be given at each post where the demand warranted and where a teaching staff could be recruited. See discussion in *Stars and Stripes*, February 28, 1919. The whole scheme was seen as "sort of a broadened Chautauqua." *Stars and Stripes*, February 14, 1919.

5. Article in *Bulletin of The A.E.F University News*, vol. 1, no. 3, May 9, 1919; *Agricultural News Letter*, publication of the Agricultural Education Section, the Army Educational Commission, number 4, April 15, 1919, in folder "Courses," Box 1939, Entry 419.

6. Article in the *Bulletin of The A.E.F University News*, vol. 1, no. 3, May 9, 1919, and discussion of the AEF agricultural clubs in the *Agricultural News Letter*, a publication of the Agricultural Education Section of the Army Educational Commission, numbers 3 and 4, April 3 and 15, 1919, in folder "Agricultural News Letters," Box 1908, Entry 408; and folder "Courses," Box 1939, Entry 419. See also *Stars and Stripes*, February 28, 1919.

7. *Stars and Stripes*, January 24, 1919.

8. Ibid., February 21, 1919.

9. *The Catalogue*, part 1, pp. 37-39; American E.F. University, Report "Registration thru April 23, 1919," in folder "Reports," Box 1965, Entry 420.

10. See Reeves's instructions to the commanding officer, sub-post at Allerey, March 5, 1919, and a list of the key military personnel at Allerey, April 22, 1919, in folder "Allerey Sub. Station," Box 1922, Entry 408. He was to sign all memoranda, orders, and bulletins as "By order of Colonel Reeves," and would carry out Reeves's orders. He could also issue his own orders and instructions, with Reeves's concurrence, for the management of the sub-station.

11. Memorandum from the adjutant general, AEF, to Colonel Reeves, March 25, 1919, folder, 188, Box 1925A, Entry 419.

12. See Memorandum, Captain Russell T. Gheen, vice principal, the Farm School, to the commanding officer, sub-post at Allerey, March 29, 1919, in folder "Details, School," Box 1940, Entry 419.

THE FARM SCHOOL

13. See a report on the school's organization dated April 26, 1919, in folder "Allerey Sub. Station," Box 1922, Entry 408, and a report of progress from the principal, the Farm School, to the inspector general, A.E.F. University, April 28, 1919, in folder "Reports," Box 1939, Entry 419.

14. See undated memorandum issued by the principal, Herbert J. Baker, outlining the organization and operations of the school farm, in folder "Establishment," Box 1939, Entry 419. Details of the faculty and its qualifications are in memorandum, Lieutenant Gilbert O. Roos, office manager at the Farm School, to Major Sandford, May 28, 1919, in folder "Rations," Box 1940, Entry 419.

15. For details of the qualifications of the faculty, see Memorandum from 1st Lieutenant Gilbert O. Roos, office manager, Farm School, to Major Sandford, May 28, 1919, and attached "List of Faculty and Personnel—Farm School," in folder "Rations," Box 1940, Entry 419.

16. *The Catalogue*, part 1, pp. 115-21. The enrollment of the Farm School at its highest was apparently 2,368 students. See Memorandum from 1st Lieutenant Gilbert O. Roos, office manager, Farm School, to Major Sandford, May 28, 1919, in folder "Rations," Box 1940, Entry 419. Report from Harry Hayward, director of the College of Agriculture at Beaune, to the educational director, American E.F. University, May 29, 1919, in folder "College of Agriculture, A.E.F.U.—Its Work and Organization," Box 1939, Entry 419.

17. See Report from Harry Hayward, director of the College of Agriculture at Beaune, to the educational director, the American E.F. University, May 29, 1919, and attached "Report on the American E.F. University Farm School" by Herbert J. Baker, principal, in folder "College of Agriculture, A.E.F.U.—Its Work and Organization," Box 1939, Entry 419. See also *The Catalogue*, part 1, p. 116.

18. Report from Harry Hayward, director, College of Agriculture at Beaune, to the educational director, American E.F. University, May 29, 1919, and attached, undated, "Report on the American E.F. University Farm School," by the principal, Herbert J. Baker, in folder "College of Agriculture, A.E.F.U.—Its Work and Organization," Box 1939, Entry 419.

19. See discussion in the unnamed, undated student newspaper, vol. 1, no. 1, folder "Meetings-Faculty," Box 1940, Entry 419.

20. See Report from the principal, the Farm School, to the inspector general, A.E.F. University, April 28, 1919, in folder "Reports," Box 1939, Entry 419.

21. Four of these exhibits, identical in composition, were created by the Department of Agriculture and shipped to France for display at various Army cantonments. This was the result of an agreement entered into between the International Committee of the YMCA and the War and Agriculture Departments and executed on February 3, 1919. Details are in a description of the "Educational Agricultural Exhibit from the U.S. Department of Agriculture," in folder "Meetings—Faculty," Box 1940, Entry 419.

22. See Report on Organization of Farm Clubs by Sergeant William H. Voigt, an instructor in rural sociology, to Herbert J. Baker, the principal of the Farm School, April 24, 1919, in folder "Reports," Box 1939, Entry 419.

23. See discussion in the *Agricultural News Letter*, of the Agricultural Education Section, the Army Educational Commission," number 5, April 29, 1919, in folder "Meetings-Faculty," Box 1940, Entry 419. See also discussion in the undated, unnamed student newspaper, (though apparently it appeared in early May 1919, and was called "The War Baby"), p. 7. See vol. 1, no. 1, of this paper, in folder "Meetings-Faculty," Box 1940, Entry 419.

24. *Agricultural News Letter*, Agricultural Education Section, the Army Educational Commission, number 5, April 29, 1919, in folder "Meetings-Faculty," Box 1940, Entry 419. The tours were especially helpful in keeping up interest in the farmers' clubs, at a time "when outdoor sports compete strongly for attention and interest," as one writer observed.

25. Report from Harry Hayward, director of the College of Agriculture at Beaune, to the educational director, American E.F. University, May 29, 1919, in folder "College of Agriculture, A.E.F.U.—Its Work and Organization," Box 1939, Entry 419.

26. There is a lengthy article on agricultural instruction in the AEF, and specifically at the Allerey School, in *A.E.F University News*, vol. 1, no. 4, May 15, 1919.

27. Several articles in *A.E.F University News*, vol. 1, nos. 5 and 6, May 22 and 30, 1919. Various clubs were also organized at Allerey as at Beaune. For example, the Masons there, numbering 175 members, were soon holding regular meetings. *A.E.F University News*, vol. 1, no. 6, May 30, 1919.

28. See copy of the *11th Regiment Bulletin*, vol. 1, no. 1, Allerey, March 21, 1919, in folder "11th Regiment Bulletin," Box 1908, Entry 408. Reeves had greeted the initial issue warmly, praising the paper's "enterprise and splendid spirit," noting approvingly that it was "full of pep." For the details on vol. 1, no. 9, see Memorandum, Colonel L. Reeves to Captain Wright, commanding officer of the 11th Provisional Regiment, April 30, 1919, ibid.

29. See memorandum to commanding officer, the Farm School, from Colonel S. Field Dallam, acting president of the American E.F. University, May 6, 1919, and indorsement of Captain Roscoe R. Snapp, Allerey, May 10, 1919, as well as a copy of the paper, in folder "Meetings-Faculty," Box 1940, Entry 419. There is no indication that the paper continued to be published.

30. Memorandum, Colonel Ira L. Reeves, to Major Patterson, executive officer at Allerey, May 21, 1919, in folder "Allerey Sub. Station," Box 1922, Entry 408.

31. Memoranda from Major Richard Brooke, district engineer, to Colonel Ira L. Reeves, May 29 and June 8, 1919, in folder "Allerey Sub. Station," Box 1922, Entry 408.

32. See "Schedule For Farmers' Institute, June 2nd and 3rd," in folder "Allerey Sub. Station," Box 1922, Entry 408.

33. Details of the fair and conference are in *A.E.F. University News*, vol. 1, no. 7, June 6, 1919. The conference was opened on the evening of June 3 with a pageant directed by William Chauncey Langdon, the university's pageant master. Dr. Butterfield delivered the opening keynote address. The conference continued from June 4 to June 7.

34. See discussions in "The War Baby," vol. 1, no. 1, and undated "Report on the American E.F. University Farm School," by the principal, Herbert J. Baker, attached to Report, Harry Hayward, director, College of Agriculture at Beaune, to the educational director, the American E.F. University, May 29, 1919, in folder "College of Agriculture, A.E.F.U.—Its Work and Organization," Box 1939, Entry 419.

CHAPTER 6:
The Art School

O UR ARMY of citizen-soldiers found itself at the end of the campaign in a foreign land which is a veritible [sic] treasure house of art of every description, whose whole history is intricately interlaced with the history of art, a land which for ages has been producing masters and master works, a land replete with museums, schools and instructors of great gifts. To our citizen artists, who are momentarily soldiers, the army authorities granted the high privilege of dropping their arms and taking up the implements of their arts. Such were the extraordinary conditions under which this school came into existence. . . .

Major George H. Gray, Commandant, Bellevue Art Training Center,
Report of the American E.F. Art Training Center, Paris, 1919

As the doughboy paper, the *Stars and Stripes*, once explained to its readers, there were "no garrets for soldier students of [the] arts in Paris." Indeed, those at the Pavillion de Bellevue enjoyed an enviable situation all around. In the first place, the view was spectacular. The pavilion, situated at Sèvres between Paris and Versailles, was on an eminence overlooking the picturesque St. Cloud and Meudon forests. It had an interesting past, and was no doubt still remembered most pleasantly by Frenchmen who, before the war, went from races at the nearby Longchamps racecourse to the popular cafe, noted for its cuisine and its wine list, that was once located there. Its history then took a different turn when it was taken over by the American dancer Isadora Duncan for use as a dance studio. It was there that she gained fame for her innovative dance forms and instructional methods. When the war came, it became a Red Cross hospital for gassed patients, before being transformed into a rather unlikely U.S. Army art school.[1] How did this come about?

When Professor John Erskine, chairman of the YMCA's Army Educational Commission, temporarily left France in the fall of 1918, in order (among other things) to select the directors of the various departments of the Army Educational Commission, he decided upon George Sidney Hellman to head the commission's Department of the Fine and

Applied Arts. Hellman was an 1899 graduate of Columbia University, from which he received the M.A. the following year. While at Columbia, he was editor of the *Columbia Literary Monthly*, the *Columbia Spectator*, the Senior Class Book, and other school publications, as well as serving as class poet and director of the athletic union. Following his graduation, he became the author of many works on art, including *Original Drawings of the Old Masters*, *Eighteenth Century Engravings*, and *Dutch and Flemish Drawings*. He was also a noted literary critic, having written on Robert Louis Stevenson and Washington Irving, among others.[2]

Flattered to be considered, Hellman first sounded out various officials, cultural leaders, and artists in the United States about the prospects of teaching art to the doughboys. He learned that not everybody in the War Department, for instance, was so sure that it was a good idea. Frederick P. Keppel, the third assistant secretary of war, for one, was skeptical. Basing his views on a short visit to France that he had taken earlier, Keppel was disappointed "in the Philistine attitude of our soldiers in France." He doubted, in fact, "whether there is a chance of teaching anything about Fine Arts" to them at all.[3] However, Hellman was informed that General Pershing was in favor of such initiatives, which, in any case, accorded with his own views that "very important results . . . [were] likely to accrue from the return to America of a group of especially qualified men, now in the American Army, who shall receive all those facilities in regard to art instruction which France is eager, and so well qualified, to give." Accepting Erskine's offer, Hellman became the director of the YMCA Army Educational Commission's Department of Fine and Applied Arts. He proceeded to develop plans to offer art instruction at several levels. First, he hoped to introduce novices to what was best in French artistic traditions. They would also receive elementary instruction in art. As to the qualified soldiers, the atelier system, which was the hallmark of the *Ecole des Beaux-Arts* system, which Hellman warmly supported, was to be the basis of organization of this level of work. "This system almost runs itself," he declared. It seemed especially adaptable to the needs of American soldier-students. Those students at an advanced level, including some who were already professionals, could lead and supervise the younger, less experienced, men in an atelier setting. They would all have their work criticized by the "patrons" of the ateliers, who would "doubtless include some of the important men in the French art world." Hellman expected good results from the program. He thought that some of the students might remain in Europe following their discharges to further their art studies. Others would bring home an appreciation for French art that would make "a permanent contribution to the cultural life of America."

Perhaps even more important, Hellman hoped that an American academy in France, similar to the established American academies in Rome and Athens, might be created. Growing out of an army atelier system, and based on French methods, it would be under American direction, and would admit American students in numbers impossible to be accommodated at the *Ecole des Beaux-Arts* in Paris. This "would be the finest kind of a 'liaison' institution between the two countries," he concluded.[4]

Hellman was soon in France, and under General Ira Reeves's direction, assumed charge of all art instruction in the AEF. Lloyd Warren, director of the Beaux Arts Institute of Design in New York City, was his chief assistant. In addition, several officers reported to him for duty. These included: Major George H. Gray, of the Engineers, in civil life an architect and city planner from Louisville, Kentucky; three architects from New York City: Captain Aymar Embury II, also of the Engineers; Lieutenant William D. Foster, another Engineer; and Ensign Archibald M. Brown of the U.S. Navy. The latter was a trustee of New York's Beaux Arts Institute of Design who became the department's purchasing agent, later serving on Bellevue's faculty. Lieutenant Howard B. Pearce, of the Air Service, a painter from Pittsburgh, and Lieutenant Charles Cellarius, of the Infantry, an architect from Cincinnati, also joined the group.

These men began the work of organizing the Fine and Applied Arts Department of the Army Educational Commission. The department was divided into four divisions: the College of Fine Arts in the AEF University at Beaune; the Hospital Section; the Art Training Center, Bellevue; and work in certain painting and architectural ateliers in Paris. A fifth area was also involved, though it was not regarded as a formal division: that of supervising art and architectural schools in such places as Le Mans, France, and Coblenz, Germany.

As plans for the American E.F. University unfolded, Hellman assisted in the founding of the College of Fine and Applied Arts, retaining the directorship for himself. The college's first associate director and dean was E. B. Homer, formerly the director of the Providence, Rhode Island, School of Architecture. The second associate director was H.B. Monges,

9. Art Training Center, Bellevue.   111-SC-153893

10. Doughboy Art Students Drawing from Casts, Bellevue.   111-SC-160973

THE ART SCHOOL

11. Students's Drawings in Architecture, Bellevue.   111-SC-160972

earlier professor of architecture at the University of California. The recruited faculty included Lorado Taft, Professor John Gaylen Howard, and Jean Hebrard, a distinguished French architect. A large group of other able instructors was drawn to a great extent from the AEF. An account of the activities and accomplishments of this college is presented elsewhere in this study.

A second venture in artistic instruction undertaken by Hellman's department involved the assigning of especially well-qualified officers and men of the AEF then stationed in Paris to study at several well-known Paris ateliers. Many of these men desired to enroll in the Paris *Beaux-Arts* atelier itself, but this was not possible. Normally, in the French tradition, there were two categories of students in the Paris *Beaux-Arts* system: the regular students, who had to pass examinations before entrance, and the informal students, who were called aspirants. These had no formal standing, received no certificates, and, while they could attend lectures and work out projects, were not allowed to work in the *Beaux-Arts* atelier itself. They could work in so-called exterior or attached ateliers. It was in this category that the AEF soldier-students were placed. They were accommodated by the Julian Academy for painting and drawing, to which seventy-five students in painting and design, and twenty-five in sculpture were assigned; the *Atelier Leloux*, which enrolled forty students in architecture; and the *Atelier Marqueste*, which took on thirty in sculpture. The men themselves had to pay monthly dues and charges of approximately thirty francs, and with the exception of easels and stools, had to purchase all of their equipment and supplies.[5]

To be sure, this study opportunity was gratefully received. Sergeant René P. Chambellan, in a letter of appreciation to Hellman, acknowledged the "boundless enthusiasm which you have inspired in me." He was extremely grateful that he had been enabled to enter the Julian Academy so that "after something like two years of abstention I was once more enabled to get back to my work." Words could not "express my joy and gratification at feeling that I was once again forging ahead," he concluded.[6]

The third division of art work under Hellman's direction was carried out by the Hospital Section of the Department of Fine and Applied Arts. This therapeutic program for patients was headed by Captain Aymar Embury II, assisted by Lieutenant Charles Cellarius. The instructors were women from the YMCA and the Red Cross, who were preferable to army

THE ART SCHOOL

12. American Students at Work, Académie Julian, Paris.   111-SC-157700.

13. French Reception, Grenoble Chamber of Commerce.   111-SC-161007

personnel in the hospital setting. About fifteen served in this capacity. By the end of March 1919, almost five hundred student-patients were enrolled in art courses in hospitals in the Paris, Bordeaux, Brest, and Savenay areas.[7]

Another area of endeavor involved the taking over or the establishment of several detached art schools, notably at Le Mans and Coblenz. The most important of these was at Le Mans, where, shortly after the armistice, the YMCA set up the AEF School of Architecture. It began its work on December 10, 1918. Though small, enrolling about twenty-five soldier-students at each session, the course was well organized and intensive. The students were placed on detached service for three weeks, and could thus devote all of their time to art instruction. The purpose of the school was to bring together men who in civilian life were either students of, or were already practicing, architecture, giving them the opportunity to study the "fine monumental buildings of [France]."

The course was quite involved, and the hours long, the school operating from 8:00 A.M. to 9:00 P.M. It concentrated on a study of French architecture "of the best epochs from the historic and monumental buildings" so much a part of the French scene. Students made actual measured drawings from the buildings themselves, plotted to scale, took extensive field notes, attended lectures, and made numerous sketches covering the history and development of French architecture. The first week was spent at Le Mans, where, by the courtesy of the mayor of the city, the students used classrooms provided in the municipal *école du dessin*. The students were also permitted to use the plaster casts from both the *école du dessin* and the city's archeological museum as models for drawings. The second week involved field work in the environs around Le Mans, emphasizing agricultural village and city planning, "with a view to putting to a practical use, in reconstruction work, the knowledge thus gained in the school." During the third week, the students were transported by army truck—and supplied with army travel rations—to such sites as Chartres, Orleans, Blois, Tours, and others in the Loire valley and elsewhere, for the study of specific buildings, notably châteaux and cathedrals. The last two days of the course were devoted to a public exhibition, at the municipal *école du dessin* in Le Mans, of the work that the students had executed during their course of study. Popularly known as the "Coxhead School," after its popular director, Ernest Coxhead, a member of the American Institute of Architects, and a YMCA secretary in

THE ART SCHOOL

the Educational Department, the Le Mans school was a model of what could be accomplished at other locations throughout the AEF. With Hellman's arrival in France, the Le Mans school was added to his other responsibilities.[8]

But undoubtedly it was the American E.F. Art Training Center at Bellevue, also under Hellman's direction, and intended for more advanced students, that was the most important of the art programs undertaken by the AEF during these months. Major George H. Gray, assisted by Lieutenant William D. Foster, of the Engineers, became the commandant of the center. Warren, the center's director of education, was soon engaged in devising the curriculum, recruiting the faculty and selecting

14. The Sorbonne, Paris.   111-SC-157285

qualified students. The new institution was to be modeled on the academies that certain nations, including the United States, had established in Athens and Rome in years before the war.[9]

Meanwhile, several prospective locations were considered for the Art Training Center. After some deliberation, Warren selected the Pavillon de Bellevue. This building was originally a hotel and restaurant built by the famed Parisian restaurateur, Paillard. Subsequently, a few years before the war, it was used by Isadora Duncan for her famed dancing school. However, in 1914, she turned the building over to the French Red Cross for a hospital. In 1918, it passed to the American Red Cross, which ran it as Base Hospital No. 6, a treatment center for gassed patients, paying a rent to Isadora Duncan of fifty thousand francs per annum. The pavillon consisted of a large building of three stories, a mezzanine, and basement; a garage and a stable; a long Adrian barracks that the Red Cross had erected; and three greenhouses. It was decided that the service detachment, soon to be ordered to the school, could be housed in the barracks, with about three hundred students comfortably accommodated in the main building. The ground floor could be transformed into classrooms and ateliers. The building seemed ideal for the uses of the center and it was speedily obtained by G-5, G.H.Q., from the Red Cross for the month their lease had yet to run. This made it possible to occupy the building at once without waiting for the time-consuming intervention of the AEF's Rents, Requisitions and Claims Department. Negotiations then ensued for a renewal of the lease, which was subsequently obtained.[10]

In the meantime, the academy's office was still "in the attic at 76 Rue du Faubourg St. Honoré," in Paris. It functioned at full tilt, making the innumerable decisions required for establishing the school, and absorbing the arrival of new personnel. Lieutenant Homer L. Chaillaux arrived from Gièvres, where he had been supply officer of the 516th Engineers, bringing quantities of supplies with him. He became Bellevue's supply officer. Others reporting included Captain Charles S. Gusman, of the Infantry, the new adjutant; Lieutenant Harry E. Reed, of the Tank Corps, the transportation and mess officer; 1st Lieutenant Robert S. Black, a Coast Artillery officer, personnel officer; and 1st Lieutenant Harold L. Leland, of the Medical Corps, who became the post surgeon.

On March 5, the office was moved from Paris to Bellevue. A few days later, on March 10, the Headquarters Detachment arrived. This unit ran the messes, drove the trucks, cleaned the buildings, and worked in the personnel offices. Among its seventy-eight men, because of the great amount of clerical work required, the proportion of non-coms was unusually high. These men were supplemented by a guard detachment numbering twenty-two. The first of the soldiers came from the 36th Division, and when they returned to their unit in late May, their replacements came from the 1st, 2nd, and 3rd Divisions. Later, civilians were employed for work in the kitchen, and as cooks for the officers' mess. The school's historian was aware that the Headquarters Detachment occupied a difficult position in that they were enlisted men doing menial work for other enlisted men, "who were thus afforded free time for their studies," but "the situation was accepted by them as true soldiers," he reported.[11]

As soon as the center was in operation, Major Gray paid a courtesy call on the mayor of Meudon, of which canton Bellevue formed a part. The mayor and Gray arranged for the installation at the school of a small handpump, part of the fire apparatus of the town. This was lent on condition that the student body would act as the fire brigade for the vicinity of the school. The town also promised to provide a garbage disposal service, though this arrangement failed to materialize, and a private firm finally contracted to haul off the garbage for its value. Gray also visited the parish priest, Monsieur l'Abbé Edelin, who agreed to rent the parish hall—the *salle du patronage*—to the school for extra space. The hall had been used previously as a hospital by both the British and the French. It was soon fitted with dark curtains and other equipment, creating a lecture hall for lantern slides and other presentations. Because it was large enough to seat the entire student body at once, it became a useful assembly hall.

Further arrangements and innovations completed the preparations of the instructional facilities, transforming the restaurant-hospital into a school. The main dining room was strung with drop-lights for an architectural drafting room. The small dining room was curtained so that nudes might pose there for the life classes. Part of the open stables was made into a painters' studio, and one end of the barracks was provided with drop-lights, creating a studio for the interior decorators. A large sculptors' studio was secured in a separate building near the school.

Living accommodations for the students were all in the main building, or the "pavillon," as it was called. Student-officers were housed two to a room, enlisted students eight to a room. Three messes were maintained: one for officers, one for student enlisted men, and a third for the headquarters detachment. The large kitchen of Paillard fame proved to be of great value, easily providing the substantial quantities of food required.

While these developments were underway, applications from prospective students were arriving in large numbers. The attached questionnaires assisted school officials in selecting the most advanced men for the Bellevue Center, while many of the others were accommodated in the more general art courses taught at the University at Beaune. Some officials were gratified to learn that the name of the first student to arrive at Bellevue was Lieutenant Eager. When he was joined on the following day by Captain Sincere, all concerned were certain of the school's success.[12] In all, 359 students reported in between March 5 and June 15, but many of the students remained for only a brief time, electing to sail home with their organizations when these embarked. Therefore, the greatest number of students at one time at the school was 268.

The school formally opened on March 24, 1919, launched by several speeches. In his remarks, Major Gray expressed the hope that an esprit de corps might be developed, making traditional military discipline unnecessary, while Warren and Hellman set forth details of the school's structure and anticipated program. For purposes of administration and discipline, the entire student body was divided into groups over which the ranking student was designated as the *massier*. Among other things, he was responsible for handing in the attendance record of his group at the end of each week. The *massier* system was employed by the French artist ateliers, where the ranking student became automatically the *massier*, with authority to appoint such *sous-massiers* as he deemed necessary. The general military formations were in charge of the senior officer student, known as the *grand massier*. All departments, groups, and tours were in charge of *massiers* and *sous-massiers*.

The school was organized into four major divisions: Architecture, Painting, Sculpture and Interior Decoration. In view of the expected short duration of the proposed term of study—about three months—and the proximity of the school to Paris and its art activities, it seemed reasonable to limit the technical work, and the study of world art, the usual emphases in art academies, and to instruct the students more generally about the state of the arts in France. Emphasis would be placed on student visits to monuments and places of artistic interest in Paris and its environs, with each division's students being directed to the châteaux, museums, and expositions featuring its specialties. As a logical corollary to this program, arrangements for the study of the French language were instituted.

Many of those on Warren's faculty were well known both in America and abroad. Captain Ernest Peixotto, of the Corps of Engineers, head of the Section of Painting, was one of the eight official artists attached to the AEF. He was the author and illustrator of many books on travel, as well as an account of his service in the U.S. Army. He had lived and studied in France for a number of years.[13] Solon Borglum, a sculptor known in the United States and abroad for his numerous monuments, especially those devoted to subjects in the American West, became head of the Sculpture Section. He was the brother of Gutzon Borglum, who later created the Mount Rushmore National Memorial, and a prominent sculptor in his own right. Solon had served in the war with French *Foyers du soldat*, being awarded the *croix de guerre* for his efforts.[14] Captain Leslie Cauldwell, of the American Red Cross, an American painter and interior designer then living in Paris, had worked during the war for the Red Cross and the *Phare de France*. He took charge of the Section of Interior Decoration and Industrial Art Design. The commandant of the center, Major Gray, also participated in the city planning courses. In civilian life he was a prominent architect in Louisville, Kentucky. He was joined in these efforts by Captain Clarence E. Howard, who in civilian life was an architect on the city-planning commission of Syracuse, New York. Another architect who served in a leadership capacity was Ensign Archibald M. Brown, who jointly headed the Section of Architecture, assisting his coadjutor, Lloyd Warren. Brown was familiar with the French scene, having graduated from the *Ecole des Beaux-Arts*. Later he established himself as an architect in New York City. Lieutenant William C. Titcomb, of the American Red Cross, was a former professor of architecture at the University of Illinois, later serving with the Red Cross in France. Another of the painters of note was Robert F. Logan, much experienced as one of the directors and instructors at the Hartford Art Society in Connecticut. These were assisted by a larger staff of lecturers, which included Americans as well as several French experts.

Because study tours were emphasized, three trucks assigned to the school went out every afternoon along a set itinerary of sites. Careful records were kept of the attendance of each student at lectures, and his participation in the work in the drafting rooms and studios.

SOLDIER-SCHOLARS

One of the innovations of the curriculum was the provision for a course of general lectures to be presented to the student body assembled together. Because the emphasis was to be on French artistic accomplishments, every effort was made to locate French experts who could lecture in English. This was not an easy task; however, ten such lecturers were located, adding greatly to the efforts of the American teaching staff.

In architecture, major presentations were made by Arthur Kingsley Porter, professor at the Yale Art School. Porter was best known for his work on the Romanesque and Gothic architecture of France and Italy. Other lecturers included John Galen Howard, professor at the University of California at Berkeley, and a French architect, J. J. Haffner.

In painting, the head of the department, Captain Ernest Peixotto, presented a survey of the French schools of art. He was followed by French experts, such as Salomon Reinach, *membre de l'Institut;* Louis Dimier, art critic for *l'Action Française;* and Louis Hourticq, an author and authority on French painting.

Major lectures devoted to French sculpture were presented by Lorado Taft, of Chicago, a well-known American sculptor.

In interior decoration, Leslie Cauldwell gave eight lectures in which he discussed the epochs and styles of French interior design.

City planning was treated by Major Gray, Cyrus W. Thomas, and the French expert, J.C.N. Forestier, who was director of parks and promenades for the city of Paris.

In the applied arts, Marquet de Vasselot, M. H. Fritsch-Estrangin, Adolphe Giraldon, and Monsieurs Demotte and Volweider held forth, assisted by J.C.N. Forestier, who spoke on landscape gardening.

Ten lectures by Emile Saillens, a professor at the *Lycée Pasteur* in Paris, on the history of France, undergirded these presentations.
The lectures were often multigraphed and distributed among the students.

In addition, the students at Bellevue had access, through the American Library Association, to nearly all the reproductions of classic works and photographs published by the houses of Guerinet and Vincent of Paris. They could also refer to the many rare books and special collections at the *Bibliothèque de l'art et de l'archéologie*. Slides for lectures were lent by the *Ligue française pour l'enseignement* and the *Musée pédagogique*.

The Department of Architecture sought to create a three-month course that would provide its students with the opportunity for technical

THE ART SCHOOL

improvement, while at the same time presenting the accomplishments of France in the field of architecture, together with "an insight into those principles of architectural design which have caused the teachings of the *Ecole des Beaux-Arts* to be sought by countless students of all countries." The department usually enrolled about a hundred students of varying levels of expertise. Following their arrival, the students were tested to determine their level of knowledge and competence. They were then classified as elementary, intermediate, or advanced students.[15]

Those at the advanced and intermediate levels worked out the current problems set forth by the *Ecole des Beaux-Arts*, though the required drawings were reduced in size so that the American students, having other classes and assignments to attend to, might complete their work. Most students completed two such problems during the course. Their projects were criticized by Victor Laloux, of the *Ecole*, and by Monsieurs Alaux and Carlu, under the supervision of Ensign Brown. The American students also visited the exposition of the *Ecole's* architecture students.

The elementary architecture students were drilled on techniques of drawing and design. All students engaged in study trips, both near Paris and to such areas as Touraine, Normandy, and Brittany. The monuments and buildings near Paris studied in detail included the *Orangerie* at Versailles, the *Pont Neuf*, the *Place des Vosges*, the *Institut de France*, the *Hôtel de Chimay*, and the *Château d'Eau* at Saint-Cloud. The local trips were supervised by 1st Lieutenant Philip L. Small, of the field artillery, an expert sketching instructor, who insisted that students master the art of pencil sketching.[16] The students were soon preparing a weekly exhibition of their sketches, which were formally criticized.

The Department of Architecture also included the Division of City Planning, headed by Major George H. Gray, Bellevue's commandant.[17] As was true generally at Bellevue, the city-planning program focused on Paris. About forty students were enrolled in the division. Several lectures were presented to the entire student body by the city planning staff, including two by Gray on the historical development of Paris. Cyrus Thomas delivered another on social and economic forces in the city's development, while J.C.N. Forestier presented lectures on Paris gardens and parks. Numerous additional lectures on various aspects of city planning were delivered. These concerned cities around the world at various times, but the emphasis was always on Paris, where many field exercises were also conducted. On a practical note, several presentations emphasized how French city planners might assist in the reconstruction of French villages and towns demolished by the war. The instructors and students were greatly aided by the availability of complete collections of official, military, and historical maps of Paris, as well as a good library of general books and lantern slides, many of the photographs being supplied by the U.S. Army's Signal Corps and the Air Service.

Captain Leslie Cauldwell headed the Department of Interior Decoration.[18] Thirty-seven students were enrolled, four of whom were sufficiently advanced to be sent to various Paris museums to work and study independently. The course work emphasized lectures and specified practical drawing problems. Weekly visits were made to châteaux and museums in and near Paris where interior decoration could be studied, such as the Museum of Decorative Arts, the Cluny, the Louvre, the Trocadero, and the Carnavalet museums, and the châteaux at Malmaison and Saint-Germain-en-Laye. Various manufacturers were visited, including those of gilt bronze, electric fixtures, hardware, carpets, tapestries, cabinets, and glassware.

The Department of Painting was headed by Captain Peixotto.[19] It enrolled over ninety students. Earlier overcrowding had led to the decision to drop about twenty of the less apt, who were given the choice of continuing their studies at Beaune or returning to their organizations. The men who remained, and others who joined later, formed a body of students that Peixotto judged "would compare favorably with those of the best art schools of Paris or New York." The students were enrolled in portrait, life, antique, landscape, and composition classes, while smaller groups took the etching and engraving classes.

The proximity of Bellevue to Paris was especially significant for the students in the painting program. They took full advantage of the reopening of museums and the holding of exhibitions at dealers' galleries, much of which had been suspended during the war years. On March 25, through the cooperation of the organization, "French Homes," and the efforts of the sculptor, Jean-René Carrière, son of the celebrated artist, Eugène Carrière, the American students made a series of visits to the studios of numerous noted French artists. These became a Wednesday afternoon institution, Peixotto remarking that "this was a privilege that no students in Paris had ever before enjoyed." In groups of about thirty students, escorted by either Carrière or Peixotto, the American students met these artists and encountered their views on art. This enabled the students "to form their own ideas as to which of the theories of present day art was best suited to their own opinions and temperaments." At Bellevue, live models posed five afternoons a week from two to six o'clock. Ninety students were enrolled in the life class, thirty-four in the portrait class, and fifty-five in the composition class. The work of these classes was formally criticized each Tuesday and Friday by Peixotto. Robert Fulton Logan did the same for the antique class, in addition to presenting a series of lectures on artistic anatomy and color theory. French artists also regularly appeared at Bellevue to lecture and to criticize the students' work. In May, a landscape class was organized by Harry B. Lachman, an American artist then living in Paris. To take full advantage of the opportunity to study in this area, students reported to the studio at 6:45 every morning, often working to dusk. The composition class worked out assigned weekly problems. The graphic arts classes—etching and engraving—remained small but intense. Louis Orr, an internationally-known etcher, conducted the etching class. Pierre Gusman, one of France's best-known engravers, directed that area of study.

About mid-May, the students at Bellevue were invited to submit designs for posters advertising the Inter-Allied Games, soon to be held in Paris. Captain Harry Townsend, one of the eight official artists of the AEF, supervised and criticized these efforts.[20] Each week the best work produced by Bellevue students in various disciplines was selected for a *concours* exhibition, which was judged by M. Cormon, a professor at the *Ecole des Beaux-Arts*.

The painting department not only stressed the mechanics and techniques of art, but made a serious attempt to stimulate the imaginations of the students, and to encourage them to think. In addition, "they were given the benefit of different . . . criticisms and encouraged ... to think out their own ideas as to what was or was not worthwhile in Art." And, "without going too far into the realms of the ideal, they were shown the difference between commercialism and high ideals in art." To these ends, the "seriousness and patient endeavor" of French artists were extolled as the ideal.

Near the end of term, two excursions capped the work of the Department of Painting. One party, led by Lachman, journeyed to Normandy, visiting artists' haunts at Vétheuil and Giverny, including a visit to Monet's home. Peixotto escorted a second group to Fontainebleau,

Barbizon, Marlotte, Montigny, Moret, and Samois, and to the homes of Millet, Rousseau, and other famous artists.

Solon Hannibal Borglum, of the Educational Corps, AEF, headed the Department of Sculpture.[21] His department established classes in life drawing, modeling, composition, and nature study, the latter stressing analogies between the anatomy of plants and other living things. Late in the term, the students, numbering about sixteen, were assigned the task of executing a large tablet dedicating Pershing Stadium, scene of the Inter-Allied Games. Sculpture students also visited many studios of French sculptors in Paris.

So important were the study trips at Bellevue, that a special Department of Paris Itineraries was created.[22] It was headed by Lieutenant William C. Titcomb, of the American Red Cross. From the outset, it was understood that an important part of the program of instruction at Bellevue was the more-or-less systematic study of the chief architectural and artistic features found in Paris and its environs. Students were at least made acquainted with the existence, location, and significance of the major sites and artistic works, a thorough study being clearly out of the question. To accomplish these ends, a systematic organization was created, dividing Paris into nine areas and planning a field trip to each area for each student. Every afternoon, a group of sixty students departed from Bellevue, sometimes by train, but usually by truck. This enabled about three hundred students to take the same trip during the course of a week. An introductory lecture, with appropriate handouts, prepared the students for each venture.

Beyond these excursions, the time from Saturday noon until Sunday night was devoted to additional trips in and around Paris. Later, the Student Council requested travel time from either Thursday noon to Sunday night, or from Saturday noon until the following Thursday night for more extended trips. The staff and faculty approved. Many trips were arranged, extending across France from Brittany and Normandy, to the Vosges Mountains, Alsace and Lorraine, and the Pyrenees to the Riviera. After each trip, students presented a written report on the artistic works they had viewed. These itineraries were administered by 1st Lieutenant James B. Carroll, of the Air Service.[23]

In order to assist in administration and discipline of the school, and to help foster an esprit de corps, a student council was created.[24] A constitution was drafted and approved, setting up an assembly composed of representatives elected by popular vote of the various departments and divisions. The *grand massier* served as president. Sub-committees were appointed to work on specific tasks. One of these helped to work out the travel itineraries. The entertainment sub-committee sought to arrange a dance, but because the student body was composed of both officers and enlisted men, separate dances would have been required under army regulations. Instead, therefore, they scheduled three concerts, all of which were "enthusiastically attended." The council was under the general supervision of Major Henry P. Sabin, an infantry officer.

Many of the students desired to study the French language and literature simultaneously with their courses. However, time was at a premium and no formal classwork in these subjects was provided. Therefore, the students organized an informal *cercle français*, called the "Cercle Entre Nous," which held weekly meetings.[25] All discourse was in French, a fine being imposed on anyone uttering any words in English. Under the direction of 1st Lieutenant Sumner M. Spaulding, of the quartermasters, the students were assisted by René M. Delamare, who often lectured on French literature, customs, and music. Molière's *Le Bourgeois Gentilhomme* was the subject of one detailed study and on one occasion, the *cercle* attended, at the *Théâtre Antoine*, a production about the life and times of Louis XIV and his court.

There were hopes that the school would continue for a second three-month course. When polled, the students overwhelmingly voted to continue the school and other artistic ventures as well. For example, of the 88 students working independently in the ateliers, 86 desired to continue. Indeed, as Hellman observed, "the enthusiastic willingness of these soldiers after months of warfare to defer their return to America, is the best indication of their appreciation of the great benefit that might accrue to them."[26] However, the accelerated pace at which the AEF was being returned to the United States made this impractical. The fixed curriculum was dropped on June 1, and travel privileges were given to those qualified until the middle of June. The closing assembly at Bellevue was held on Sunday, June 15. The commandant addressed the students, as did Warren, and the heads of departments and the chief *massier* gave shorter, extemporaneous speeches. Immediately afterwards, about 150 of the students left directly for Brest and their ships home. Others received a ten-day leave, following which they also embarked at Brest. The men attached to organizations that were part of the Army of Occupation in

Germany returned to the Rhine. A few of the men elected to take their discharges in Europe, many continuing their studies at various schools there.

There is little doubt that Bellevue was a remarkable success. Discipline problems were few; most of the students were well aware of the valuable opportunity that they enjoyed. At the highest levels, officers also appreciated that "for an army to undertake to conduct an art school . . . in the field," was an unprecedented operation. In the first place, it became policy to give all students the same military status during their studies, while at other times observing the usual, familiar distinctions between officers and enlisted men, making provisions for their messing, sleeping, and living apart. This common status in class was necessary to avoid duplication of courses. The commandant noted in his final report that "from the scholastic point of view this worked out most satisfactorily; from the point of view of military discipline, it resulted in an apparently insurmountable tendency to laxness as to military courtesies." But Gray felt no need to take time from studies to correct these military defects; on the contrary, he thought "that the less there were of military restraints in evidence, the more would the student acquire of that freedom of imagination and unhampered mental attitude necessary to any artistic receptiveness and conception." In short, he concluded, "it was essential . . . to develop an artistic atmosphere." To assist him and his staff in achieving good conduct "without obtrusive military restrictions," the students were encouraged to govern themselves and, to the extent possible, put on their honor. They were informed that as select men, they would be granted every reasonable privilege, but were they to abuse it, they would lose it.

Certainly the soldier-students fully appreciated their opportunities. Their views were perhaps best stated by a contingent from the Department of Sculpture, which appended a letter to the final report submitted by Borglum.[27] They praised the daily lectures, which helped them, "with our young and comparatively simple Western minds, overflowing with life and action," better to understand the richer, older, and more complicated, civilization of France. The unheard-of opportunities to enter the ateliers of celebrated French artists had been most valuable, as were the numerous excursions and longer trips. The give-and-take of discussion and criticism of their work was likewise appreciated. "In all, it has been a great profit to have had the opportunity of associating with so many fine, intelligent young fellows, so very able in their own arts, so strong in vitality, possessors of varied experience and firm in their convictions." Finally, they wanted to be remembered "as a class of hard earnest workers who have striven to derive the utmost from the opportunity which has been presented them." Perhaps a painting student said it best: his three months at Bellevue "had made up for the loss of two years work while in the Army."

One forgets that, in that far-off time, high-minded civilians and military personnel seized the opportunity to ease the negative aspects of the war experience by turning military facilities and the norms of service to positive account. If the world had been saved for democracy, then some of the fruits of that victory could already be harvested. The reverse contact of New World back to Old could be used to their mutual benefit, though, no doubt, the American doughboy-student was the main beneficiary. The French, who assisted unstintingly, certainly thought in terms of enlightening the Americans who had so recently given of their blood and treasure without stint. Then, too, the French were no doubt flattered to be asked to expound upon their culture and civilization, something they are generally—and understandably—prepared to do with much pride. The so-called "lost generation" of Americans who came to Paris in the 1920s is certainly misnamed. It might better be perceived as truly a "found" generation: where would Hemingway, and so many of the others, have been without their Paris years? Some of the first trickles of the later surging stream of Americans pouring into France were impelled by the AEF's educational innovations.

Yet Hellman hoped for much more, indeed, for nothing less than the beginning of a significant American art tradition. He was greatly encouraged by the unexpected response to the art program among soldiers of the AEF. He was impressed "that . . . in an army of young Americans, there should be thousands so lured by the vision of beauty as to wish to make its practice their life work." America lacked a true artistic tradition because it was a young, raw nation, he declared, but such "traditions must begin somewhere, and what time is better than this to begin our own traditions of art?" This was in general a time of "great beginnings," he added, "and our army can bring no more inspiring message to our nation than this, that the art impulse is strong and enthusiastic among the youth of America." Moreover, he was confident that America was destined to "wrest from Europe the leadership in the realm of art; for Europe is old and weary and we are fresh and young." To be sure, a sympathetic public was required. Hellman was certain that this would emerge as well. One of the positive things that the war had done for Americans was to bring them into fruitful contact with the artistic traditions of Europe. Therefore, "we are henceforth done with the old belief that art is a thing apart, a frill, a mere ornament of life—something almost effeminate." In fact, the art programs of the AEF had revealed "that hands which held firmly the bloody bayonet are now eager to grasp the architect's T-square, the painter's palette, [and] the chisel of the sculptor." Americans no longer needed to "hark back to the days of Pericles, or learn from warrior kings and dukes of Italy and of France, that art is a masculine force to be encouraged and cherished by fearless peoples; our own officers and enlisted men are teaching us this truth."

Yet there was also the hope that art could contribute to the sense of fellowship among nations, as was already often the case among individuals, with permanent salutary effects. In fact, art was "the sole universal language," Hellman insisted, "and as the appreciation of beauty in the scheme of existence grows among mankind, war itself will become more and more remote."[28]

Hellman's perennial hope for the establishment of an American art academy was eventually realized. The Bellevue ideal was not allowed to die. Even in its short life, it had left an indelible impression upon its students and faculty, many of whom reflected at its closing: "What a pity that some similar institution could not be founded upon a more permanent basis!"[29] Consequently, in 1923, a wing of the Palace at Fontainebleau, southeast of Paris, was transformed into the Fontainebleau School of the Fine Arts. There, with aid from the French government, advanced American art students could study for three months in a program based on the Bellevue model. At the palace, the school joined the Conservatory of Music for American Students, which had been established two years previously. The music school was an outgrowth of another AEF institution, the Bandmasters School, located at the headquarters of the AEF in Chaumont. It was under the direction of Walter Johannes Damrosch, well known on the New York music scene before the war.[30]

The art school had as its goal, not to study the technique of art, but "more particularly to awaken the intelligence of its students to the more serious problems of art and stimulate their interest in the relation of the various plastic arts to each other." Based on the methods perfected at

THE ART SCHOOL

Bellevue, the emphasis would be on study tours to public and private art collections, to the sites of major buildings, and to the studios of great artists. The general director of the Fontainebleau School of the Fine Arts was M. Laloux, the famous *patron* of the *Ecole des Beaux-Arts* in Paris, who had been the mentor of numerous well-known American architects. The American Committee was concerned with the recruiting of the students who would be involved. Its architecture department was headed by Whitney Warren, of the Beaux Arts Institute of Design in New York City. The department of painting and sculpture was headed by Ernest Peixotto, who had directed the department of painting at Bellevue.[31]

SOLDIER-SCHOLARS

NOTES

1. There is a copy of the lease agreement between Isadora Duncan and the American Red Cross to be in effect from June 18, 1918 to three months following the end of hostilities. The Red Cross in effect sub-leased Bellevue to the art school. See in folder "A.E.F. University. Historical Data," Box 1956, Entry 409. See also *Stars and Stripes*, May 2, 1919.

2. Report, on the Department of the Fine Arts, Army Educational Commission, Y.M.C.A., December 10, 1918, in folder "Physics," Box 1926, Entry 419. It should be noted that it was Hellman who insisted upon the department's being called the "Department of Fine and Applied Arts." It had originally been styled only the Department of Fine Arts.

3. Letter, Hellman to Erskine, New York City, November 15, 1918, in folder "Director of Fine Arts," Box 1924, Entry 408, and pamphlet "Army Educational Commission, Y.M.C.A., Department of the Fine Arts," November 20, 1918, in folder "Reports," Box 1925, Entry 419; and letters, Keppel to Hellman, War Department, Washington, December 4, 1918; Hellman to Keppel, New York, November 30, 1918; and Hellman to Erskine, New York City, November 15, 1918, in folder "Director of Fine Arts," Box 1924, Entry 408.

4. See discussion in Report of the Department of the Fine Arts, Army Educational Commission, Y.M.C.A., by Hellman, December 10, 1918, in folder "Physics," Box 1926, Entry 419, and a printed pamphlet, containing an introductory note regarding the "Army Educational Commission Y.M.C.A., Department of the Fine Arts," New York City, November 20, 1918, in folder "Reports," Box 1925, Entry 419.

5. See memorandum, from George S. Hellman, to Colonel Charles W. Exton, Paris, February 5, 1919; attached unsigned description of the Paris *Beaux-Arts* system, Paris, February 15, 1919; and one from C. W. Thomas to a Mr. Homer, undated, all in folder "Paris; Ateliers, Beaux-Arts, Museums," Box 1926, Entry 419. Colonel Charles W. Exton was a GHQ, Fifth Section, officer stationed in Paris as General Rees's representative there.

6. See his letter of February 24, 1919, in folder "Letters from various sources to Mr. Geo. Hellman," Box 1925, Entry 419.

7. See memorandum, G. S. Hellman to Grosvenor Atterbury, March 28, 1919 in folder "Letters from Geo. S. Hellman," Box 1925, Entry 419, and undated progress report, Captain Aymar Embury II to Hellman, and report Embury to Hellman, Paris, February 22, 1919, with attached Memo, February 21, 1919, in folders "Public Hygiene," and "Hospital Work," Box 1925, Entry 419.

8. See letter, Ernest Coxhead to John Erskine, Le Mans, January 27, 1919; unsigned report on the "School of Architecture Program," Le Mans, December 24, 1918; a detailed itinerary for the School of Architecture, Le Mans, for the three-week period, January 27 to February 16, 1919; a clipping from the Le Mans newspaper, *La Sarthe*, January 28, 1919, devoted to the school; a small collection of snapshots of the students and their activities, in folder "Coxhead School," Box 1925, Entry 419. See also Mayo, "That Damn Y," pp. 307-9 for details as to the founding of the school at Le Mans.

9. The details of the Art Center are largely derived from: U.S. Army, *Report of the American E.F. Art Training Center, Bellevue, Seine-et-Oise, March-June, 1919* (Paris:

THE ART SCHOOL

Frazier-Soye, 1919), especially Major George H. Gray, "Report of the Commandant," pp. 9-12; Lloyd Warren, director of education and dean of the faculty, "Report of the Director of Education," pp. 19-25; 1st Lieutenant William D. Foster, official historian, "A Narrative of Events," pp. 87-93, and various individual reports by the heads of the departments and divisions. Other reports from some of these men are cited below.

10. These details are in undated Report on the American E.F. Art Training Center, in folder, No. 195, Box 1925A, Entry 419.

11. Foster, "A Narrative of Events," p. 92.

12. First Lieutenant Robert J. Eager was from the Quartermaster Corps and was enrolled in the architecture course. Captain Edwin M. Sincere was in the Engineers, enrolled in the architecture and city-planning course.

13. See Alfred Emile Cornebise, *Art from the Trenches. America's Uniformed Artists in World War I* (College Station: Texas A&M University Press, 1991), pp. 19-21, 50-52, for a discussion of Peixotto's career. His work on the AEF was *The American Front* (New York: Charles Scribner's Sons, 1919). His travel books, all published by Charles Scribner's Sons of New York, include: *By Italian Seas* (1906), *Romantic California* (1910), *Pacific Shores from Panama* (1913), *Our Hispanic Southwest* (1916), and *A Revolutionary Pilgrimage* (1917).

14. Solon Borglum died unexpectedly on January 31, 1922, at the age of fifty-two. Gutzon commemorated him in an article, "Solon H. Borglum," *American Magazine of Art*, 13.11 (November 1922), 471-75.

15. Lloyd Warren, "Report of the Department of Architecture," pp. 29-31, 36-50.

16. See 1st Lieutenant Philip L. Small, "Report on the Course in Pencil Sketching," pp. 34-35.

17. See Major George H. Gray, "Report of Division of City-Planning," pp. 31-33.

18. Leslie Cauldwell, American Red Cross, "Report of the Department of Interior Decoration," 51-52.

19. Capt. Ernest Peixotto, "Report of the Department of Painting," 57-60. The French were impressed with what Peixotto's presence meant in France, creating him a chevalier of the Légion d'honneur in 1921, and an officer of the legion in 1924.

20. For Townsend, see Alfred E. Cornebise, ed., *War Diary of a Combat Artist, Captain Harry Everett Townsend* (Niwot: University Press of Colorado, 1991).

21. Solon Hannibal Borglum, Educational Corps, AEF, "Report of the Department of Sculpture," pp. 73-74.

22. Lieutenant William C. Titcomb, American Red Cross, "Report of Department of Paris Itineraries," 79-84.

23. See "Report on Itineraries of the More Distant Trips in France," pp. 110-12.

24. Report by Major Henry P. Sabin, "The Student Council," 109.

25. First Lieutenant Sumner M. Spaulding, Quartermaster Corps, "Cercle Entre Nous," 113.

26. Memoranda, George Hellman to General R. I. Rees, May 12 and 14, 1919, in folder "College of Fine Arts and Applied Arts, etc. Correspondence 1 to 11, etc.," Box 1924, Entry 408, and Memorandum, Major George H. Gray to General R. I. Rees, May 8, 1919, in untitled folder, Box 1914, Entry 419.

27. See letter appended to the "Report of the Department of Sculpture," pp. 74-75.

28. See published address on the "Applied Arts and Education," which Hellman

delivered on April 8, 1919, at an educational conference at the university at Beaune in untitled folder, Box 1914, Entry 419. The address was published as Bulletin Number 93 for the university by R. De Thorey of Dijon on May 1, 1919.

    29. "The Fontainebleau School of the Fine Arts," *American Magazine of Art*, 14.2 (February 1923), 84.

    30. The records of Damrosch's school are in RG-120, Records of the American Expeditionary Forces (World War I), 1917-23, Entries 421-423, "Chaumont: Bandmasters and Musicians School."

    31. See article, "The Fontainebleau School of the Fine Arts," *American Magazine of Art*, 14.2 (February 1923), 84-87.

CHAPTER 7:
Relations with the French

RELATIONS between the Americans and the French were always problematical. Pershing himself had furious arguments with the French high command, especially regarding the necessity, as he saw it, that the Americans create their own force. The French and the British had originally wanted the Americans to integrate their troops with the Allied armies as reinforcements. Pershing held out for a unified, separate American command, which he was eventually able to create. And with over two million Americans on French soil, there were naturally clashes and conflicts. These intensified after the armistice. The Americans awaiting their transports home were often bored and intensely irked by contacts with the French, especially by what they regarded as gouging French merchants. In return, the French were unhappy to find so many Americans still in their midst. Instances of street fighting occurred in St. Nazaire and in other cities, and in Nice, a gang ambushed and killed several American military police.[1]

Yet there were those at the university who insisted that Americans should understand that "worn and disabled as in many ways she now is, France is still the country toward which artists and thinkers love best to turn, as likely to find there intelligence, reasonableness, and sensitiveness to beauty, and a clear choice always for the things which make the soul of man great."[2]

Nevertheless, the relations between the men at the university and the city of Beaune were predictably mixed. Some matters that surfaced were trivial, others more substantial. Some prohibitions and limitations vis-à-vis citizens, whose homeland the men shared, came into play. Ideals often gave way to a mundane, sometimes harsh, reality. The French *commissaire de police* at Beaune sent instructions that no motor vehicles driven through the streets of the city must exceed ten miles per hour. Drivers were also cautioned that during rainy weather, when the streets were slushy, they must use care so as not to splash pedestrians or "crowd them off into the mud."[3] Soldier-students on Sunday excursions were warned about trespassing and the need to maintain amicable relations with the French.[4] The army's bus route between the campus and Beaune was only for the use of officers and civilians associated with the university, so enlisted men went on foot to town. When they did so there were some limitations: they were instructed not to go or to return to Beaune via the railroad yards, and to avoid friction with railway authorities.[5]

Much of the contact between Americans and Frenchmen was in the sphere of commerce. Certain regulations and restrictions applied. French price controls were in effect, which, with American cooperation, helped to keep excessive pricing in check. The office of the *sous-préfet* in Beaune reminded the newly-arriving Americans of a decree of June 30 and an order of September 26, 1918, that had been instituted by the local French government when the Hospital Base was constructed. These specified that prices for food, toilet articles, clothing, and the services of hairdressers must be prominently displayed in English as well as in French. The *sous-préfet*'s wishes, "together with those of the population of Beaune," were that no merchants should be allowed "to take advantage of either the generosity or the ignorance of the American soldiers." To be sure, the American authorities were, at the same time, to impress upon their own personnel that they were not to purchase any article on an attached list if its price was not properly posted, or if merchants asked for a price higher than that stipulated.[6] And those officers obtaining permission to live in Beaune were strictly forbidden to pay more than fifteen francs per day, room and board, or one hundred francs per month for the room alone.[7]

The Americans were reminded in turn that they must not run up accounts with any French store, restaurant, or business establishment. All purchases were to be in cash and in accordance with the official price list.[8] Furthermore, because the area of Beaune was essentially a wine-producing country, poultry and dairy products were "extremely scarce." With the large influx of Americans in Beaune, the supplies could not meet the needs of both the townspeople and the doughboys. Accordingly, American personnel were asked, "insofar as possible," to refrain from buying eggs and milk, "thus saving the townspeople of Beaune discomfort."[9]

There were problems pertaining to unhealthful conditions. Because the water supply of Beaune, and of most of the towns and villages in the vicinity, was declared unsafe by American health officials, new sources had to be tapped on the base.[10] A scarlet fever epidemic broke out in the towns of Ladoix and Serrigny, which were placed out-of-bounds for

American troops.[11] From time to time, French establishments were also placed off limits to American personnel, as was the Café Deserteaux, for instance, for a month from February 16 to March 17, 1919.[12] An establishment owned by Charles Barthelmy, at the junction of Milan and University roads, was similarly proscribed because of unsanitary conditions.[13]

But there was much more evidence of cooperation and goodwill. The main hotel in Beaune, the Hôtel de la Poste, became a popular gathering spot for Americans. It permanently set aside a large private dining room for officers, YMCA, Red Cross workers, and nurses. On Wednesday and Saturday nights, dinner dances were scheduled there.[14] Also of significance in Beaune was the city's Committee of French Homes, of the Association of French Homes. This national French organization had been created to extend the hospitality of French families to the members of the AEF.[15] The Administrative Council of the Beaune Committee also established an officers' club in the city.[16]

The French-American Club, under the auspices of the *sous-préfet* of Beaune, maintained an office to assist students enrolled in French courses at the university, whether enlisted men or officers, though a special room was reserved for the latter.[17] American personnel were invited from time to time to concerts held at the *Foyer du soldat*, some of which were sponsored by the organization, *Les amis de la musique*, of Beaune.[18] The city's museum offered privileges to the university. Though it was open only on Sundays, instructors could arrange to hold their classes there at other times, taking advantage of the museum's substantial holdings in Egyptian bronzes and other antiquities, the substantial mineralogical collection, and the good collection of representative art of French and Flemish schools. Many items that had been on display at the *Exposition Universelle* in Paris in 1878 were retained by the museum as well.[19]

Benefits to relieve what the French had suffered were common after the war. The Americans were kept informed of the pressing needs, with frequent requests for assistance. One specialized organization, the Association of Public Utility, held numerous benefits, often at huts of the *Foyers du soldat*. The association, with chapters in each department and arrondissement of France, was empowered to raise money for a certain class of war invalid. During the war, France had been compelled to recruit many underage and physically weak men for military service. Many of these were not strong enough to bear the strenuous demands made upon them, and fell victim to diseases, especially tuberculosis.

Declared unfit for further service, these men were discharged and sent home without pensions, unlike those disabled by wounds. The Association of Public Utility was created to assist these unfortunate former soldiers, who were unable to work, and often needed special treatment. As was common in those days, the receptions that they held featured flowers, cakes, and other refreshments usually provided by the ladies of communities, and often concerts presented by "young artists of the best families of the city." A typical benefit, scheduled in Beaune on the afternoon of March 30, "cordially invited" men of the university.[20]

The Anti-tuberculosis Society of Beaune was also active in raising funds. And the Committee of Devastated Areas of France was another far-reaching, active organization that sponsored concerts and charity fairs, the proceeds going for the relief of the areas ravaged during the war.[21]

The Americans responded to French appeals and also repaid French hospitality with musical performances. For instance, in the Beaune *Parc de la Bouzaize* on Sunday afternoon, May 25, the university presented a well-attended concert that drew much praise.[22]

On Mothers' Day, May 11, the French and Americans staged a joint program. Festivities began with a concert by the College of Music students and the university band, performed in the English Gardens in the city.[23] In return, the French held receptions and in other ways helped the Americans to celebrate the occasion, though they had heretofore been unfamiliar with the holiday. Indeed, their participation was greatly appreciated by the Americans, and "no more delicate, no more appreciated compliment could have been tendered us," one news writer in the student newspaper said.[24]

More elaborate was the program arranged for Thursday, May 29, when a large French delegation, headed by the *ministre de l'instruction publique* and other high officials, inspected the university. A full-scale military review of all the provisional student regiments was held in the late afternoon on Pershing Field, followed by other festivities and ceremonies.[25]

On the following day, touching the still sensitive heartstrings of both the French and the Americans, an elaborate, colorful joint Memorial Day ceremony was held. Featuring military and religious programs, the results clearly brought the townspeople and the university community closer together.[26]

Whatever the relations were elsewhere in France between Americans and Frenchmen, those between the soldier-students at the AEF

University and the citizens of Beaune were, on balance, of a positive nature. This state of affairs was commented upon by Reeves, as the university's career drew to a close: "It is believed," one of his memoranda began, "that the relation of the Americans with the citizens of Beaune has been as cordial as that of any station in France." He praised officials and citizens of the city for their most hearty cooperation in all of the school's endeavors, which had made for a most pleasant stay. It seemed only appropriate that the personnel of the university leave behind some substantial monument, funded by voluntary subscription and designed by the school's fine arts department.[27] The structure was intended to commemorate French soldiers from Beaune who had died in the war and Americans who had died in area hospitals. Both funds and plans for the memorial were to be left in the hands of the mayor of Beaune "for him to use as he sees fit." A copper box that would contain the names of the donors was to be set in the base of the monument. And though, as Erskine was also well aware, "we will not get to see this memorial completed," in years to come, "when we come back to Beaune, as we all will," it would remain "as a reminder of the university as we know it now," and no doubt of the close relations that existed between Americans and Frenchmen during the months of the university's existence.[28]

In the event, personnel at the university raised 11,712.10 francs for the memorial. This was subsequently erected in Beaune, though it was dedicated only to the "*Enfants de Beaune, morts pour la Patrie.*" Colonel Reeves was made one of the *présidents d'honneur* of a committee for planning the creation of the monument. There is no indication that the plans of the fine arts department were considered in designing the structure.[29]

Much later, on June 24, 1979, the city of Beaune unveiled a plaque on Docteur-Tassin street in the former American camp, recognizing that an American field hospital and later the AEF University had existed in that old quarter of Beaune.[30] In 1986, a hut that had originally been constructed by the Americans, also located on Docteur-Tassin street, was restored and turned into a joint Franco-American museum dedicated to the collection and display of materials from the World War I era. At the same time, from November 8 to November 29, 1986, the municipal archives of Beaune mounted a large exhibit recounting the history of the American camp.[31]

Clearly, the citizens of Beaune still keep alive memories of both the American hospital and the university, recognizing their impact on the city's history.

NOTES

1. Franco-American relations are discussed in Frank Costigliola, *Awkward Dominion. American Political, Economic, and Cultural Relations with Europe, 1919-1933* (Ithaca: Cornell University Press, 1984), 169-72. There is also information regarding French attitudes toward Americans at Beaune in Perriaux, *Le Camp Américain*, pp. 25-28, and passim.

2. Address by John Erskine, May 20, 1919, published by the University as Bulletin No. 100: *French Ideals and American* (Dijon: R. De Thorey, 1919).

3. Memorandum No. 11, Headquarters, American E.F. University, March 19, 1919, Entry 415.

4. Memorandum No. 6, Headquarters, American E.F. University, March 13, 1919, Entry 415.

5. Bulletin No. 5, Headquarters, American E.F. University, February 21, 1919, Entry 414.

6. Bulletin No. 6, Headquarters, American E.F. University, February 28, 1919, Entry 414.

7. Memorandum No. 5, Headquarters, American E.F. University, March 10, 1919, Entry 415.

8. General Orders No. 9, Headquarters, American E.F. University, February 26, 1919, Entry 412.

9. Memorandum No. 20, Headquarters, American E.F. University, March 30, 1919, Entry 415.

10. Memorandum No. 54, Headquarters, American E.F. University, May 12,1919, Entry 415; Bulletin No. 45, Headquarters, American E.F. University, March 31, 1919, Entry 414.

11. Memorandum No. 52, Headquarters, American E.F. University, May 10, 1919, Entry 415.

12. Bulletin No. 5, Headquarters, American E.F. University, February 21, 1919, Entry 414.

13. Memorandum No. 67, Headquarters, American E.F. University, May 28, 1919, Entry 415.

14. Bulletin No. 24, Headquarters, American E.F. University, March 17, 1919, Entry 414. The Hôtel de la Poste was also the terminal of the base's bus service.

15. See *Bulletin des French Homes* (Paris), number 4, February, 1919, which featured Beaune's *comité*, the president of which was M. Grimanelli, the *sous-préfet* of Beaune. See in folder "French," Box 1926, Entry 419. The *comité* also raised funds by canvassing. See notice in the *Journal de Beaune*, April 23, 1919, that up to April 18, 1919, the city had raised 615 francs for the effort. See copy of this paper in folder "Investigation by I.G. and Other Officers," Box 1922, Entry 408.

16. Bulletin No. 35, Headquarters, American E.F. University, March 26, 1919, Entry 414. This club was at No. 24, rue Paradis, Beaune.

17. Bulletin No. 47, Headquarters, American E.F. University, April 2, 1919, Entry 414.

18. Bulletin No. 49, Headquarters, American E.F. University, April 3, 1919, Entry 414.

19. Bulletin No. 54, Headquarters, American E.F. University, April 8, 1919, Entry 414.

20. Bulletin No. 38, Headquarters, American E.F. University, March 28, 1919,

RELATIONS WITH THE FRENCH

Entry 414.

21. Bulletins Nos. 80, 112, and 115, Headquarters, American E.F.University, May 6, 27 and 30, 1919, Entry 414; Memorandum No. 63, May 23, 1919, Entry 415.

22. Bulletin No. 105, Headquarters, American E.F. University, May 23, 1919, Entry 414.

23. Bulletin No. 83, Headquarters, American E.F. University, May 7, 1919, Entry 414.

24. Newspapers in France sought to educate their readers about the significance of Mothers' Day in America. See, for example, article in *Journal de Beaune*, April 23, 1919, in folder "Investigation by I.G. and Other Officers," Box 1922, Entry 408, and article in *A.E.F. University News*, vol. 1, no. 4, May 15, 1919.

25. Memorandum No. 66, Headquarters, American E.F. University, May 27, 1919, Entry 415; Bulletin No. 113, Headquarters, American E.F. University, May 27, 1919, Entry 414. Detailed article in *A.E.F. University News*, vol. 1, no. 6, May 30, 1919.

26. Memorandum No. 68, Headquarters, American E.F. University, May 29, 1919, Entry 415. Detailed accounts in *A.E.F. University News*, vol. 1, no. 7, 6 June 1919.

27. Memorandum No. 60, Headquarters, American E.F. University, May 20, 1919, Entry 415. See also stories in *A.E.F. University News*, vol. 1, nos. 5 and 7, May 22 and June 6, 1919.

28. Article in *A.E.F. University News*, vol. 1, no. 5, May 22, 1919.

29. See letter, June 12, 1919, from the mayor of Beaune to Reeves, in the municipal archives of Beaune, collection 7 I § 15 art. 8, and other relevant documents also in the same collection, and in M I § 15 art. 8.

30. See account in the local paper, *Le bien public*, May 9, 1989.

31. Accounts ibid., November 8/9 and 10/11, 1986; May 13, 1992.

CHAPTER 8:
Other Universities

AMONG THE most fortunate of the American soldier-students were undoubtedly those who attended a college or university in Great Britain. Dr. George MacLean, former president of the University of Iowa and subsequently associated with the YMCA and the Army Educational Commission, who then became secretary of the London branch of the American University Union, was appointed by the Army Educational Commission to take charge of the assignment of men to the British universities.[1] The first detachment of soldier-students arrived in the British Isles on March 4, 1919, at Knotty Ash Camp, near Liverpool. There, for the first few weeks, the men spent time "worrying around in the snow and rain, watching bulletin boards, filling out blanks and what not." Once things were sorted out, however, they became "almost a part of the British population." Eventually assigned to their respective schools, the students were also issued passes permitting them to roam freely about the country, and they spent much time in sightseeing and travel.[2]

All together, 1,051 enlisted men and 972 officers were involved, becoming widely dispersed to colleges, universities, and other academic and research institutions throughout England, Scotland, Wales, and Ireland. The largest contingent was at the University of Edinburgh, with 237 officers and men, closely followed by the 229 enrolled in London University's famed School of Economics. Cambridge accommodated 192 students, while Oxford's detachment numbered 159. Other institutions enrolling significant numbers included the Universities of Aberdeen, Birmingham, Manchester, Glasgow, and Dublin; the University of London's University College, King's College, and its Fellowship of Medicine; and the University of Wales at Aberystwyth. The College of Legal Education in London, as well as other law schools and legal institutions in the capital, such as the Inns of Court, enrolled over 200 students. Others accommodating somewhat fewer numbers included the Universities of Bristol, Belfast, Liverpool, and Sheffield, and Birkbeck College in London. Four enthusiastic agricultural students, all enlisted men, studied dairying and dairy farming at the University of Reading, while four other agricultural students were assigned to the Rothamsted Experimental Station at Harpenden, about forty miles north of London. In all, forty-nine institutions accepted one or more American soldier-students, usually for something over a three-month period of instruction and research.[3]

The American students represented all of the states of the union. Most had studied a year or more in some American college or university. Many already possessed degrees and took graduate work in Great Britain. Some had studied in the British Isles previously, notably several Rhodes scholars, a number of whom returned to Oxford as part of that university's detachment. Of these, a few took Oxford degrees. For instance, Sergeant W. C. Bosworth, who had been awarded his B.A. Oxon. previously, was made a Master of Arts, as was Captain Homer L. Bruce, the adjutant of the student detachment at Oxford, who had taken his B.A. at Oxford in 1915. In addition, Captain John V. Ray, was at last able to pick up his B.A.[4] Another former Rhodes scholar, Lieutenant Laurence A. Crosby, of the student detachment there, became a member of the teaching staff. He was asked to take over the duties of a Trinity College law tutor still absent in the British army. Well qualified for his responsibilities, Crosby was a graduate of Bowdoin College, and held law degrees from Oxford and Columbia.[5]

The student detachments in the British Isles were commanded by Colonel F. F. Longley, whose headquarters were in Russell Square, London. For administrative purposes, he divided the United Kingdom into three districts, and assigned commanding officers to each of the institutional detachments. In certain instances, as at Queen's University, Belfast, University College, Reading, and University College, Galway, the commanding officers were sergeants. In others, only one student was enrolled, as was the case with Lieutenant E. S. Turpin at the Royal College of Music in London, and Captain H. M. Black, at the Royal Normal College for the Blind.[6]

Most of the commanding officers seem to have ruled with a light hand, and they and their student charges worked together in easy harmony, making the stay in England a pleasant and profitable one.[7] The degree of work that was done varied considerably, however. In some instances, the men arrived at their schools just as the spring holidays began. Such was the case with the couple hundred American soldier-lawyers who arrived at the moment when the Inns of Court were closing for nearly a month's vacation. The new term was to begin on June 18, but the men were to start homeward by the end of the month, leaving little time for serious work.[8] The Birmingham *Mail* once observed that the seventy-four soldier-students at Birmingham University had not "been overworked in Edmund Street," but that probably the real object of their stay was to learn something about Birmingham's businessmen and their methods, with a view to applying the knowledge to the expansion of international trade in the future. To these ends, they had been enrolled in special courses in mining and commerce, and had been frequently entertained at several highly successful "smoking concerts"—a new term added to the Americans' vocabulary—arranged in their honor by the city's Rotary and Cosmopolitan clubs. The doughboys reciprocated by introducing Birmingham "to the joys of Jazz and the excitements of baseball."[9] One frank, if perhaps naive, doughboy-student, in his required semi-monthly report, stated that "during the last two weeks I have attended teas, dinners, banquets, theater parties and dances."[10] Nevertheless, much serious work was done. The students enrolled at Birkbeck College embarked upon a special course in economics, beginning on March 27, which continued at the rate of two hours daily throughout the term, with the exception of one week at Easter. A general survey of the whole subject of economics was accomplished before the course was completed.[11] Similarly, the four agricultural students enrolled at University College, Reading, eagerly pursued a systematic course of study regarding dairy farming, learning such things as cheesemaking. They also journeyed to the Jersey and Guernsey Islands to observe dairying methods on the home grounds of famous dairy herds.[12]

To be sure, travel and sight-seeing remained central to the educational experiences of the American soldier-students, and went on incessantly. The men took full advantage of passes that permitted them virtually unlimited travel privileges throughout the British Isles. They were often assisted in their travels by the Y staff at the famous "Eagle Hut" in London.

One site visited by many of the men was Stratford-on-Avon, no less popular than with generations of tourists before them. Especially noteworthy was attendance during "Shakespeare Week," at which time many of the American students participated in the numerous ceremonies and festivities of the occasion.[13] Shortly after Shakespeare Week, Secretary of the Navy Josephus Daniels placed a wreath on the grave of Shakespeare in the spirit of Anglo-American fellowship, a watchword during the time of the American students' sojourn.[14] On Memorial Day, many of the students journeyed to American grave sites where the British often joined the Americans honoring their war dead.[15]

Typically, two students hitched a ride to the Scilly Isles on a tramp steamer; some journeyed to the Isle of Skye; yet others bought bicycles and toured the whole of Scotland. The Killarney Lakes in Ireland were a popular destination, as was Blarney Castle, near Cork, Ireland, where obligatory trips were taken to view the Blarney Stone. One group of students from Aberystwyth reported that they had even caught Colonel Longley, "hanging head downward," kissing the famous relic.[16]

A major theme often discussed during these weeks was Anglo-American solidarity and amity. Various organizations fostered this. One of the most closely involved was the English-Speaking Union, an organization headed by the former president of the United States, William Howard Taft, and Arthur James Balfour, the British foreign secretary. It was established—significantly—on July 4, 1918, and had as its purpose the drawing together of the English-speaking peoples of the world, women as well as men, though with an emphasis on British-American relations. It published a magazine, *The Landmark*, and maintained an information bureau, a club, and a library in London for the convenience and assistance of its membership. Its goals were also embraced by the London branch of the American University Union.[17] The American Club, which also promoted a better understanding with the British, came into existence at Oxford. A similar club was organized at Edinburgh University.[18] Another Oxford organization was the British-American Club, with an even clearer goal of promoting understanding and friendly relations between the British Commonwealth and the United States.[19] Members of the American soldier detachment who were attached to the U.S. Army Medical Corps were especially active in exchanges and contacts with their British counterparts. They organized the American Medical Post-Graduate Society in London, with the expressed goal of fostering scientific and social relations between English and American members of the medical profession. Similar aims were advanced by the Anglo-American Committee of the Fellowship of Medicine.[20]

There were other significant contacts between the British and the American soldier-students. Numerous organizations, as well as private persons, sought to interact with the members of the American student detachment. Many American residents in Great Britain opened their homes to their visiting compatriots. Most notable among these was Mrs. Waldorf Astor, who entertained hundreds of students in numerous "at homes" at her mansion at 4, St. James Square in London.[21]

One of the chief attributes of the American soldier-students, both in England and in France, was their great interest in athletics, especially as practiced on American college and university campuses. Soon various team sports emerged, with particular emphasis on baseball. Where units were too small for such an organization, tennis squads flourished. Many Americans also captured places on rowing teams, some being included on the prestigious crews at Oxford and Cambridge. More informally, the Americans enjoyed punting, with predictable results: "Several of our enthusiastic students have already been immersed in the river while pursuing this adventurous sport."[22]

Attempting to explain baseball to the British and in turn learning the intricacies of cricket[23] was a mutually significant encounter. Indeed, as one writer declared, it was "one of the great results of the sojourn of the Americans at Oxford." The British were certain that the American pastime was "a nasty rough game," while Americans were equally sure that cricket was "a slow, uninteresting, long-drawn-out misery." But both revised their original estimates of the others' respective sports, admitting "that they were badly mistaken in their first impressions." Nevertheless, as one observer noted, it was not to be seriously expected that either nation would adopt the chief sporting event of the other, "for after all, are not the two games but reflections of the national temperaments of two people, and equally a manifestation of that love of outdoor sport which characterises the English speaking race from all other races of men?" Also, "the games are an expression of the life of two great countries, slowly evolved through decades out of different conditions. Each nation knows its own game, and knowing it loves it better."[24]

The American students at Manchester, especially, went to great lengths to make baseball popular there. The sports editors of the local papers were given detailed briefings on the game, and spectators were presented with folders containing information about the sport. In addition, it seemed that "any American who has not taken an English lady to a baseball game has missed a delightful afternoon," one newspaper correspondent asserted. "Their knowledge of their own sports has been a pleasant surprise, and they have proved apt in understanding the main features of baseball."[25]

But if the British were not convinced of the superiority of the American sport, the American students were able to leave England, "with a certain knowledge that the American student detachment has been instrumental in teaching the English the rudiments of a great American game," with some bold prophecies that "before very long our dignified and reserved cousins" would adopt baseball and become as avid and vocal fans as were Americans.[26]

At the same time, notably at Birmingham, several of the students seriously attempted to learn "all about the game of cricket."[27] At both Oxford and Edinburgh, other students turned to golf, though the caretakers were not impressed, one noting that "after the students have finished the course it looks as if a farmer had just done his spring ploughing."[28]

Nevertheless, the national pastime remained the most important of the sports the students pursued, and eventually eleven baseball teams were fielded by the various detachments, Oxford's being, by all accounts, the best.[29]

Simultaneously, as one newspaper report observed, while the American students at Bristol were "playing baseball, boating, cycling, and dancing," they were also "promenading on the Downs." And it was whispered "that the last named activity [had] attracted a one hundred per cent. turn-out."[30] Undoubtedly, the promenading, not to mention the numerous dances, and other social engagements, led to significant meetings between many British lasses and soldiers. The results were predictable: numerous marriages occurred in the relatively short time that the Americans were in the British Isles. Indeed, "that doughty little archer," Cupid, was "still piling up munitions, despite the armistice, to proceed further with his campaign of entangling alliances between the male citizenry of the U.S. and the gentler sex of England, Ireland, Scotland, and Wales." About ten marriages, for example occurred within the detachment at University College, London, out of the 120 students there. In Edinburgh, there were nine reported among the about 260 students enrolled. But, for some reason, none occurred at Oxford.[31]

The American stay in England was also memorable because even the weather cooperated. For over forty days there was uninterrupted sunshine, surely a "boon that nature has provided for the American students in England." They were not unaware of their good fortune, one writer observing that England had put on its "prettiest summer frock for us to see with happy eyes." The American students would certainly long remember that "perfect May and ... inimitable June."[32]

But these good things came to an end, and with considerable anticipation. One soldier-poet at Oxford, Douglas Lawson, captured something of this spirit:[33]

SOLDIER-SCHOLARS

Soon Home!

Can't you feel it in the air;
It's around you everywhere,
Lilting like a lover's tune:
"Home soon, Home soon!"
You hear it in the lecture hall
Like a siren-voice's call;
Telling wanderers who roam:
"Soon home, Soon home!"

On the river; in the hills;
Everywhere you go it thrills
With its song of budding June,
"Home soon, Home soon!"
. . . .
It leads us forth from England's shore
Like Pied Piper led of yore,
Chanting from the ocean's foam,
"Soon home, Soon home!"

An anonymous poem similarly caught the spirit:[34]

Finale

We're changing our pounds and our
    shillings,
  Our sixpences, farthings, and such;
For it's not many days till, God will-
    ing,
  We'll be where we'd rather be—
    much.
. . . .

We're running like mad o'er the
    Islands,
  To see all we can while we can—
From Cork to the top of the Highlands,
  From Blackpool to Southend—oh,
    Man!
. . . .
It's all nearly ended, however,
And soon we'll be catching the boat.

OTHER UNIVERSITIES

> "Believe me when we say we'll never
> Leave mother again"—end quote.

Nevertheless, it had been a scintillating experience, if not "an epoch in education," as George E. MacLean alleged, concluding that "as the Crusaders to the Holy Land . . . inaugurated a new epoch for Europe, so may we ... for America." He hoped that the men might return home "bearing the winged torch of knowledge and sympathy, to keep alight the new spirit of international opportunity and obligation in place of the old spirit of selfish isolation."[35] An editorial in the *American Soldier-Student,* No. 7, June 25, 1919, warmly concurred, noting that "no academic world ever flung open its doors with greater hospitality than did this island one." But the classroom and lecture hall were not the only sources of educational experiences: "We made the whole kingdom our campus." Indeed, "never perhaps in the history of the world has a body of students travelled so much in a limited period of time as we did. We made the so-called 'wandering students' of the Middle Ages look like hermits." What could be done in return? Possibly the men could best be a power "in the realm of Anglo-American relationships" if all could help Americans at home "see and understand the British people as we do now. If we do this, the interesting educational experiment we are now completing will have proved to be a success, even if not a single one of us ever 'cracked a book.'"

Perhaps that perennial anonymous poet best summed it up, in his poem, "A Toast":[36]

> Britain, you've been good to us;
> > On the square.
> We are grateful for the fuss;
> > Put it there.
> You have used us "jolly fine,"
> And your girls have been sublime;
> Hope we meet another time,
> > Over there.
> They had told us you were cold;
> > 'Tisn't true.
> Said that you would call us bold,
> > Even rude.
> But your spirit, as of old,
> Was to us as true as gold;
> Yes, you took us in your fold;
> > Here's to you.

SOLDIER-SCHOLARS

By mid-June the soldiers began to leave, the first being twenty-man contingent from University College, London, departing to Liverpool, bound for Brest, France, and thence to America. Others followed shortly after the end of the lectures at their respective schools, usually between June 20 and June 30. Most of the men had embarked from the United Kingdom by early July. The only disappointed members of the Student Detachment belonged to the United States Third Army, the force that was engaged in occupation duty along the Rhine. These had to return to their German stations, one soldier dolefully advertising in the student paper that he would gladly exchange "one visit to Germany and two service stripes for one fourth-class passage to the United States."[37]

More numerous were the doughboy-students enrolled in fourteen of the major French universities, as well as other schools. These accepted 7,500 American students into their hallowed halls. The Sorbonne headed the list with over 2,000 students, followed by the University at Toulouse, where 1,200 men were accommodated. The University of Montpellier admitted 559, the University of Grenoble accepted 371, 373 were enrolled at Lyon, 249 at Poitiers, 204 studied at the University of Aix-Marseille, and 147 matriculated at Besançon. Other universities involved were Caen, Rennes, Nancy, Dijon, Clermont-Ferrand, and Bordeaux.

The origins of the involvement of American soldier-students with French universities dated to December 1917, when French educational officials contacted the American University Union about the possibility of various French universities' accepting a limited number of qualified American soldier-students for enrollment when time and circumstances permitted. This offer was dispatched to the YMCA, which designated A. P. Stokes, chairman of the board of trustees of the American University Union, to organize such a program. Stokes was soon succeeded by Professor John Erskine, who became head of the Y's Army Educational Commission; he further developed the program. Soon American students were being processed and enrolled.[38] Therefore, American soldier-students were involved even before the armistice, and following the end of hostilities, by the spring terms, the program of admitting American doughboys to courses of study in French universities was well advanced.

Inevitably, as American soldier-students were reminded, there were considerable differences between the French and American higher educational establishments. To go from one to the other would require "patience, intelligence and adaptability on the part of the student."

15. University of Montpellier.   111-SC-160793

Students should not expect the French to turn their universities into American colleges, though they made every effort to accommodate the doughboy-student. The European university in general "makes no effort to be encyclopedic in its curriculum; these did not attempt to cover the whole field of human knowledge," one source explained. Courses were given by specialists in their own field and many important subjects "in a given year, even at the University of Paris, are not treated." Students should therefore content themselves with the work that was offered. It would, of course, be impossible to conduct advanced courses in English in all subjects that were being taken by the French students.

Courses of two sorts were offered in all of the universities: public lectures (*cours publics*), for which there were no fees and to which no formality of attendance or examination was attached, and closed courses (*conférences*) which required matriculation and an interview with the professor. The latter were conducted much like classes in American institutions. The language was a problem, and accordingly, the *Alliance française*—in Paris it was allied with the Sorbonne—instituted a series of French courses, at various levels, designed to integrate students as rapidly as possible into the regular university lecture halls.

In addition, the French went to considerable lengths to establish special courses for American students at the various faculties of the Sorbonne, including those of medicine, letters, law, and sciences. Other institutions made similar arrangements, such as the *Ecole des hautes études sociales*, the *Ecole supérieure pratique de commerce et d'industrie de Paris*, and the *Ecole supérieure d'électricité*, where a special course in wireless telegraphy was set up. For those interested in science, the *Institut Pasteur* admitted a number of researchers into its laboratories, while the *Observatoire de Paris* established a special course in astronomy. Other Paris schools in commerce, political science, and agriculture also made provisions for Americans in separate classes, as did a number of technical institutions and certain faculties of theology.[39]

Naturally, the programs varied, but the men were usually admitted to the established courses and schools, with law and letters predominating, though many were also enrolled in engineering, science, and medicine. Specialized studies were not ignored. At the University of Lyon, for example, five students were enrolled in technical courses pertaining to the silk and dye industry, while six more studied with the veterinarians at the Lyon city animal hospitals.

Perhaps typical were the students at the University of Montpellier. There, of the 449 men in the Law and Letters school, 220 took regular French courses, including sculpture and painting, while 229 opted for special concentrated courses in French to develop their proficiency before they enrolled in the regular faculties at the University.[40]

It was customary, as at the University of Poitiers, for the students to obtain diplomas at the end of their courses to certify their attendance at the school. This was in addition to the certificates in teaching and in French literature given to more advanced students. Students admitted to the course in advanced French were required to pass a preliminary examination. Only then could they seek a diploma testifying to their ability to teach French "in foreign countries." At Poitiers, for example, the students admitted to the diploma program met every Tuesday and Saturday throughout the course, diligently preparing for their written and oral exams scheduled for the close of the semester in late June.[41]

Erskine, who was in charge of all of the soldier-students both in the British Isles and in France, was more closely involved with the latter. General Rees allowed him to nominate the officers to command the detachments in France, and he worked through the Yale, Harvard, Princeton and Columbia bureaus at the University Union in Paris, seeking the best men available to direct the work. He succeeded in locating first-rate officers. For example, he placed Captain Robb in charge at Dijon. A Yale man, he was "simply ideal," Erskine observed. Not only was he efficient as a commander, but he had "completely won the hearts of the French—whether or not he has persuaded them to take our attitude toward college life."[42]

Some of the units were organized along military lines, with provisions for military drill. However, in some instances, the students were given a freer rein. At many of the universities, student councils assisted the commandant in administration to a considerable degree. For example, they sometimes checked attendance in classes, which was not part of the French system. All soldiers in the AEF were ordered by GHQ to put in six hours of work per day, with the soldier-students undertaking three hours of serious study. An honor system was instituted whereby students were required to certify that they spent this minimum time daily in academic work.[43]

Nonetheless, each of the detachments was separate, and all operated in rather different fashions. At the University of Toulouse, a large one with over twelve hundred students, a substantial organization was established to run their affairs. The Business Council of the American Detachment at Toulouse University was formed on March 7, 1919. It received and disbursed all moneys from the students for tuition, board, lodging, and other assessments. The United States Army paid commutation of rations and paid for the quarters for the students, though apparently the tuition costs were borne by the students themselves. The council also exercised a general supervision of all student service agencies, and executed all relevant contracts. In addition, it sought to foster and promote the growth of literary, dramatic, journalistic, and athletic organizations at the university. To these ends, three officers and seven enlisted men served on the council and worked in conjunction with the commandant and the student body.[44]

The detachment at the twin universities of Aix and Marseilles was treated as one unit despite the twenty-five miles separating the campuses. The commandant was Captain Kenneth F. Simpson, assisted by a supply officer, an adjutant and twenty-one soldiers, including members of the Medical Corps. Regarded as "an extremely capable young officer, of pleasing personality," Simpson spoke French, and was "in every way fully capable of conducting the University activities." He had a smoothly-operating office. The American dean of the detachment was Professor Baillot, a French scholar. Students enrolled in sciences, medicine, and engineering were taught at Marseilles, while those studying law and letters were at Aix. About 90 percent of the students lived with French families, the balance in hotels, villas, and pensions.[45]

The unit at the University of Montpellier consisted of 142 officers and 454 men. They were commanded by Captain Sherley W. Morgan, assisted by a supply officer, an adjutant, and 16 men, including four medical personnel. Lieutenant Colonel Gordon R. Catts, the inspecting officer of G-5, felt that Morgan was "a particularly capable and efficient officer, [who] speaks French fluently, [and] possesses great character and personality."[46]

The American soldier-students in France were much more concentrated and focused on their work than were the men in Great Britain. In the first place, they did not possess the liberal travel passes that their fellows in the United Kingdom had. GHQ later provided all soldier-students with passes that permitted those in good standing to range around the university center up to ten kilometers. This was hardly comparable, however, to the system in Great Britain.[47] Nevertheless, the soldier-students in France enjoyed much more freedom than their peers in the traditional military units. They seemed to have been blessed with considerate commanding officers, almost all of whom understood that the university setting was not to be confused with the usual cantonment. The soldier-students were greatly assisted in exploiting their windfall, the opportunity to study in famous old institutions of higher learning in France. Inspecting officers found few disciplinary problems, though some students were sent back to their organizations, a possibility that was a deterrent to most of the others. The AEF's concern with venereal diseases led to frequent inspections with generally satisfactory results. The French officials and professors were almost unanimous in praising the American students, noting especially their seriousness and studiousness.[48]

Something of their aims was addressed by Major H. H. Vreeland, the commandant of the detachment at the University of Bordeaux, in an interview with a reporter of *Les Beaux Jours*: "We in Bordeaux feel that it is our duty primarily to learn the French language, the French literature, the French history, but above all, the manners and customs of the French people, among [whom] we live."[49] Dr. John Charles Dawson, president of Howard College at Birmingham, Alabama, civilian dean of the detachment at the University of Toulouse, agreed, noting that the key to the involvement of the American students was to seek Franco-American cooperation and understanding.[50] Or as a soldier-journalist in Montpellier's student paper, the *Soldier-Student*, perhaps best put it,

16. Trinity College, Dublin, and Doughboy Students.   111-SC-161586

17. Soldier-Student Detachment, University of Bristol.   111-SC-161058

18. Student Council, University of Lyon Detachment.   111-SC-161016

19. Track Team, University of Lyon Detachment.   111-SC-161023

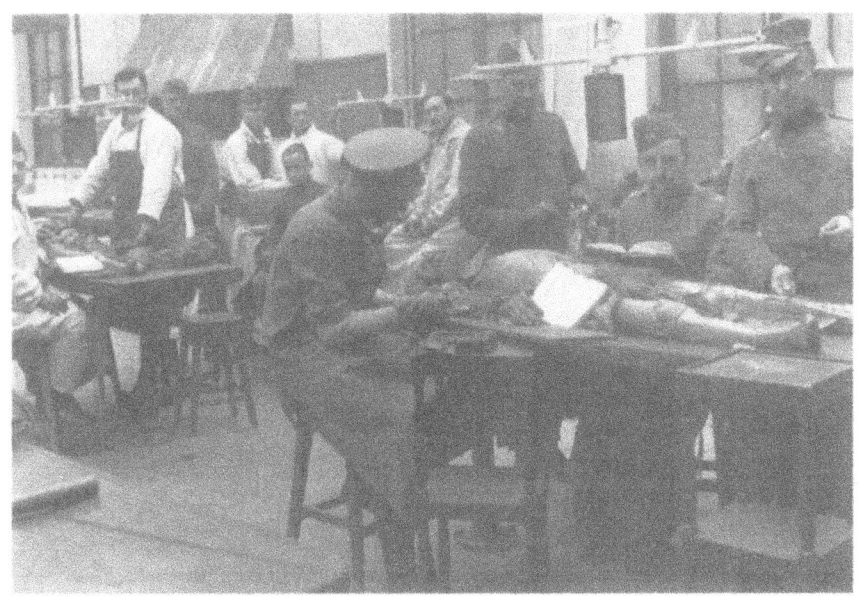

20. Dissecting Room, University of Lyon.   111-SC-161012

21. Outdoor Class in French, University of Grenoble.   111-SC-160999

22. Dining Hall, University of Toulouse.   111-SC-160786

the Army student detachment there had arrived "to learn and read the inner soul of that people, whose cause had become their own."[51]

The men were well aware of the rare opportunity they had been given. Certain aspects of that experience were captured by one A. Buchanan in his poem, "More Strange than Fiction":

> More strange than fiction; truth, my lad,
>   You've often heard it said.
> Little ye wot toward fair Toulouse
>   The path of duty led.
> When first you heard your Country's call
>   To stop the hordes of Huns,
> Little then you thought of books and pens
>   But much of swords and guns.
> You did your bit, the chevron gold,
>   Perhaps a wound stripe, too,
> Describes more eloquently than words
>   The Hell you have been through.
> But that's all past and shoved behind.
>   At present day we meet
> At Culture's shrine, our face incline,
>   And sit at Wisdom's feet.
> For some, the classic, sylvan dales
>   Of Poetry, Music, Art,

OTHER UNIVERSITIES

> While others seek the plainer road
> > Mapped by Academic Chart.
> And some will delve in Legal Lore
> > Of Rights divine, humane,
> Knowing full well 'twill never be
> > The old Regime again.
>
> And some will seek, with vision high,
> > The Truth, whate'er it be,
> From depth of earth to starry sky
> > They may the clearer see.
> Truth? It may be our studies here
> > Will prove the forge, perchance,
> That welds with links of Love sincere,
> > America and France.[52]

    One of the consequences of this flourishing academic venture was a reassessment of the men's basic philosophies and ways of life. In addition, there was much speculation about the differences between American and French cultures, and about how much Americans perhaps had to learn. One of those who became interested was Lieutenant Charles B. Parmer, who was particularly impressed by the differences in musical tastes. "In America rag time is the music of the masses, and opera the costly diversion of the classes," he asserted, while in France opera was "a portion of the knowledge of peasant, bourgeoisie, and noblesse." France was indeed the "Olympus of artists," a situation most unlike that of America. In the United States, the pace of history had been too fast, the struggle for the "Almighty Dollar" paramount. In addition, in certain prominent religious sects, "out of touch with the aesthetic," music was regarded as immoral. Accordingly, "there was no place for the dreamer of dreams in our pre-war existence. We [lived] life swiftly, so our national music partook of the spirit of the times." He hoped that now "that we have conquered the forest, subdued the savage, made the land one of plenty, and have taken our place in the Seats of the Mighty," Americans might devote a little time and thought to the arts: "We loudly boast that our plumbing is the best in the world but who wishes to dream of sewers? Now that the war is over, let us turn our gaze toward the stars!" Among the men most strategically situated for this endeavor were the soldier-students attending French and British universities.[53]

    Similarly, regarding art, the doughboy-student was encountering new, refreshing insights. This amounted to nothing less than a revolution, because "in matters of art in every form our artists have been compelled to trail their dreams in the dust of mock-morality." In fact, since colonial times, one writer observed, "our workers in fine arts have been fettered to earth and their wings legally clipped by those who could not trust themselves, and therefore would not permit their neighbors to gaze upon representations of the 'human form divine' sans fig-leaves." But America seemed at last to be "emerging from its Chinese wall of provincialism," and the American students in the French universities were catalysts in the process. They were having a full opportunity to view impartially the highest arts of "the mother of civilization," and would be interpreters of advanced art to those at home who had had no chance to study abroad. The students were viewing art, hearing operas, and reading the French classics, reserving for themselves the right to determine "whether art, *sans fig-leaves*, is immoral." Certainly, "art and mock-morality in the States are not congenial. Their matrimonial bounds were forced upon them. We suggest a divorce."[54]

Routinely, articles and editorials in the student press discussed the quality of the opportunity that the men had and what they might be expected to gain as students abroad. If they were to obtain the maximum from their exposure to French life, they must have an open mind about the way the French viewed and lived life. The articles frequently revealed the larger horizons and changed attitudes that the European experience had produced in the hearts and minds of the doughboys. Now especially detested were the so-called "professional uplifters" who made it their business to appear as the guardians of the nation's morals. The doughboys felt that many of these were responsible for Prohibition and much else that the majority of the men in khaki now perceived as detrimental to a decent, humane life. The doughboys understood that they had earned their right to resist what others at home might foist upon them: "They tell us things aren't quite the same back in the old U.S.A., and some of us who have been over here for the last two years know we aren't quite the same," one editor noted. "We know a lot more about work, the blessedness of sleep, about death and the fear of death. Then, we know there are others who have learned a great deal more about those things than we have—French, by the way, and English. . . ." Therefore, "when the Professional Uplifter demands that we be protected from the contamination of Continental Ideals, we smile."[55]

If the men were contemptuous of the attitudes of such uplifters, the politicians at home were likewise warned that the two million members of the AEF would be a political force to reckon with when the men returned home. No longer would those holding their seats "through sheer oratory," as had been true for so long, find themselves unchallenged; they would be confronted with a power in politics "which they cannot stem with mere wordy flights into the azure blue." A new standard would be required; politicians who did their duty and took the business of politics seriously would receive the support of the men of the AEF, who had learned much about duty and efficiency. Therefore, "a frock coat will not hide a faker from us in the future."[56]

One perennial problem involved what many in America regarded as the immorality of the French. Lieutenant Franklin F. Massey of the United States Army Medical Corps addressed this matter, lauding the French for their frankness, which stripped off the veneer of unreality. The French recognized passions, emotions, and desires as being integral parts of human nature and took them as a part of everyday life. This attitude was manifested in their art, for example, and the French "frankly admire the human form and nudity," which was to them no more thought of as being indecent than was a carbon drawing of a woodland. As to clothing, the French had their own styles, which also were regarded in certain circles as being too scanty or too revealing of the human form beneath. But this was another manifestation of frankness, not of licentiousness. In France, it is true, he went on, that sinfulness and the chaste exist side-by-side. But surely America would do well to examine its own habits and customs, and consider them with the same degree of clear-cut, frank criticism that the French do regarding their own.[57]

Considerable furor was aroused by the publication of an article in the Paris edition of the Chicago *Tribune* which reported that a certain Miss Elizabeth Marbury, in the United States, had strongly criticized the roles and actions of the women workers serving the AEF, especially those in the service organizations. Marbury had feared the close contact between these women and the soldiers, asserting that "human nature is the same on the Marne as on Broadway." The negative reaction was vocal across France in various troop papers of the day.[58]

More serious was an article in *Collier's Weekly* by Mark Sullivan, "one of America's foremost publicists," urging that the United States immediately withdraw the boys from Europe before they adopted "the French attitude toward women, and the Continental tolerance of alcohol." To doughboy students at Toulouse, this was an insult. To be sure, the men wanted to come home, but in the meantime, some found themselves in French universities, and in a foreign country, "one which has filled the pages of history with immortal glory." While there, the men were determined to observe its life and customs. But Americans at home need not fear that the doughboys would become less "American" in the process: "Our backbone isn't that supple ...; we are not babes in swaddling clothes," they insisted. It was certain that the French had an attitude toward women "which is different from ours." However, that was their affair: "our job was to help lick the Hun, not to attempt to 'reform' the Continent." The average American had a far too high regard for American women "ever to hold them in a lesser light than our own mothers taught us."[59] Indeed, "few American soldiers found permanently attractive what were assumed to be French standards of sexual promiscuity." And, many of them disliked France, at least in comparison with the United States. They found the French dirty and the country and language strange, while the women the soldiers met were generally prostitutes. Moreover, "that France was associated with the Army and the war was enough to rob it of much of its charm for bored, homesick men."[60] To be sure, the Americans who met or lived with French families developed considerable respect and even love for them, but in general, wartime France was a difficult place to admire, except for what the French had endured. As to marriages, a few thousand did occur throughout the AEF, the Army's attitude being "cool but officially neutral toward the idea of Franco-American marriages."[61] Nevertheless, "the men who knew her best were content to leave Mademoiselle in France. They would sleep with her if they got the chance, or most of them would, but few of them wanted to marry her or teach her ways to American girls." Home remained the doughboy's primary goal, where he could "see 'an honest-to-God American girl' again."[62]

Mother's Day seemed an appropriate time to elaborate on several concerns being expressed in the United States about the boys overseas: how they had changed, what the war had done to them, and how they regarded their mothers and the American home. Dorothy Anne Skinker, of the YMCA staff in Toulouse, in an open letter "To The Women At Home," published in the student paper *Qu'est-ce Que C'est?* attempted to answer. To her, there were certain results worthy of comment: "*The soldier has seen life very simply and for that reason very clearly,*" she declared. He had faced a harsh reality and was accordingly "impatient of sham and hypocrisy." Indeed, "he has an insight into character that is quite disconcerting," she went on. "In everything he 'has to be shown.'" This was especially true in the field of religion. Yet this was not a development to be deplored, she continued. If he succeeded in carrying "out this keenness of insight into civil life, I believe he will have a much finer, firmer basis upon which to build his life." As to his devotion to his home, the soldier constantly dreamed of it, longed for it, and idealized it. Home had become for him a shrine and a talisman. All, therefore, seemed well; there was little to worry about, she concluded.[63]

Clerics and churchmen in America were especially targeted by thoughtful spokesmen arising among the soldier-scholars of the AEF. The old, narrow morality was not enough for the combat veteran. The stresses and strains of the battle experience had led the soldiers to a simple philosophy and theology. The soldier in line "lives each day as it comes, regards sleep and food among the great boons of life, gages his superiors accurately, if cynically, and snaps to attention at the voice of authority." He did his job stoically, cursing fluently when the occasion demanded, sought the elemental joys of life "whole-heartedly and unblushingly," all the while, remaining "a profound lover of justice," with a tender spot for all those down on their luck. He was a harsh judge of his officers, choosing those who knew their job above all. He had a primitive faith in the justice of God, and was little interested in complicated theological systems. His basic philosophy was "live and let live." All of his activity, "all this fatigue, these marches and counter-marches, this vast uncertainty of life and experience with fear, [had] left him a curiously primitive creature." He respected the church, but its complexity merely disturbed and confused him. Religion he continued to believe in, "but organized religion [had] fallen short of his ideal of leadership."[64] To be sure, the doughboys had to make decisions as to how they would spend their Sabbaths, the French being known for their "Continental Sundays," emphasizing family gatherings and purely secular entertainments and outings. Many Americans readily took to "a bit of harmless recreation on the Seventh Day, especially as we are soon to return home where the Commandments are literally enforced." However, not all were so inclined, at least one soldier observing that many American soldiers abroad "[were] extremely cautious in their assumption of foreign attitudes towards institutions so closely related to our national vitality and moral inheritance as the American Sunday," preferring, indeed, "the Sabbath of hallowed tradition."[65]

But if the American soldier-students spent time in rethinking their deepest attitudes and philosophies, and developed a new appreciation for the British and French, they hardly surrendered their own life-styles, especially clinging to the attributes and characteristics of their former academic life. They almost immediately set out to reconstruct the traditional collegiate organizations, institutions and practices at the

French and British universities. One familiar aspect of traditional American campus life was the debating societies, clubs, and teams that were soon duplicated in France. An AEF College Debating League was begun, involving the student detachments at Poitiers, Toulouse, Lyon, Grenoble, the Sorbonne, and the AEF University at Beaune. Its fortunes were closely followed in the various student newspapers that also sprang to life, as in *Les Beaux Jours*, at the University of Poitiers, for example.[66] The debaters at Poitiers beat out Toulouse and the Sorbonne to win the championship in France, and prepared to meet the champions of Great Britain in a later match.[67] Debating teams were often formed by members of the various divisions represented on the student detachments. One seemed particularly well set up to produce verbal fireworks: members of the 1st Division challenged members of the 42nd Division to debate the subject, "Resolved—the military record of the 1st Division in the war between the United States and Germany shows that its performance out-ranks that of the 42nd Division." More traditional topics were later debated by teams of the various universities, organized into a league by the YMCA, as, for example, "Resolved, that there should be a minimum wage law enacted in the United States by appropriate legislation."[68]

Another activity that attracted attention was the establishment of detachment newspapers, sometimes by journalism students, though not always. The official organ of the American students at the University of Montpellier was the *Soldier-Student*, a four-page weekly published every Saturday in conjunction with *Le petit méridional*, the local *journal républicain quotidien*. Two of the pages were in English; the remaining two were in French. *Les Beaux Jours* served both American and French students at the University of Poitiers. This eight-page weekly also appeared on Saturdays. *Voilà!* was the weekly of eight pages published by the students at the University of Bordeaux. *Qu'est-ce Que C'est?* a sterling, clever paper, was undoubtedly the best of those of the student detachments in France. The twelve-page weekly, including a lively sports page, appeared every Wednesday. The Americans at the University of Rennes published *As You Were*. With its size and the resources at its disposal, the student detachment at the Sorbonne might have been expected to publish an outstanding paper. However, much to the disgust of *Les Beaux Jours*, the Paris students used the *New York Herald*'s Paris edition as their paper. The Poitiers students wanted to know if there was not enough support and cooperation among the Sorbonne students to produce a representative publication. They should publish a paper of their own, *Les Beaux Jours* chided, "one that is characteristic of [their] spirit," rather than shift that responsibility to a commercial paper.[69] *Deux Mots* was the publication of the students at Clermont-Ferrand University; the *Lorraine Sentinel* served the men at the University of Nancy; the *Besançonian* appeared at the University of Besançon. Grenoble published the *Alpine-American*, while the University of Dijon's students produced the *American Dijonnais*.

Apparently, though, there was little inclination to permit the establishment of the Greek social fraternities among the students; however, former members of the fraternities from schools in the United States did organize and socialize informally.[70]

And other groups were formed. At Toulouse, fifteen men interested in Christian work and the ministry met with the chaplain to organize a discussion group for the exchange of ideas upon the role of the church as illuminated by the experience of the men in the war. They also hoped to pursue courses at the Protestant Theological Seminary at nearby Montauban, which five soldier-students later did, spending the last two months of their stay at the university there.[71] At Montpellier, the chaplain also organized a Students' Association made up of soldiers desiring to pursue or resume preparation for the ministry and missionary work, as well as social service work. This group subsequently held weekly meetings.[72]

In fact, "veritable epidemics" of organizations occurred at the universities. At Toulouse, the Faculty of Letters Franco-American Club was organized; the law students created their club, as did the Beaux Arts men. There was even an "Empire State Club of Toulouse University," as well as another organization composed of men from Texas and Oklahoma. An Alabama Club was later formed. One specialized group was the International Trade Relations Society of the World organized at Toulouse, which brought in speakers to address the students. One was Dr. James T. Shotwell, the Columbia University historian who was serving as chief of the History Division of the U.S. Commission to Negotiate Peace in Paris. He discussed the drafting of the Treaty of Peace and the League of Nations.[73] Another interesting organization was the "*Entre Nous* Club," composed of fifty French and American men and fifty French women, all students in the Faculty of Letters, University of Toulouse. To foster better French-American relations, the Franco-American Club was organized, every student in the university being a de facto member.

23. Lecture Hall, University of Dijon.   111-SC-161030

24. Industrial Lab, University of Dijon.   111-SC-161011

OTHER UNIVERSITIES

At Poitiers, a new student society, the "Order of the Bolsheviks" appeared, though it had little to do with politics or political philosophy. The students made one M. Girard, the proprietor of the Café Castille, their honorary president, and the officers were given titles of nobility, i.e., "Baron de Cohn, Viscount of Kansas." The order's name was chosen because it was one "which every one knew, but which no one understood." In addition, it was a name "not easily forgetable, no matter what the physical state of the brother," while retaining "all the horrible elements of the unknown."[74]

Some observers cautioned against the student organizations. In the view of the editor of *Les Beaux Jours* at the University of Poitiers, cliques and groups, more interested in themselves and in their group interests, than in the detachment or the university as a whole, had formed. These had created a "deplorable state of lethargy," which he attacked in an editorial in the paper. He strongly appealed to all the men to seek rather "to exemplify the spirit that we can find in any university in the States." "Let us," he insisted, "in the organization of our work and play, be first and last, American!"[75]

As was the case in England, the American soldier-students formed athletic programs and took great pride in introducing their sports to the natives, taking considerable pains to describe and demonstrate baseball, as well as football, and even pushball, to local French citizens.[76] Some of the squads went on tour as, for example, the baseball teams of the universities of Marseilles and Aix. They journeyed to the island of Corsica and staged a baseball game in the town square of Ajaccio, "not 200 yards from where the 'Little Corporal' [Napoleon] was born in 1769." There, American corporals got base hits. The teams were those who got seasick, and those who did not. The game was won by the "rail birds" by a score of eleven to nine, thus beating their "stronger- stomached bunkies." The victory enabled the "rail birds" to proclaim themselves the baseball champions of Corsica, as well as of all of the islands of the Mediterranean.[77]

But not all of the American soldier-students were interested in athletics. Music and the arts also attracted large followings. Many of the students took advantage of the opera season, which they could enjoy in all of the university centers. Many were surprised at the rich operatic and theatrical fare that was available to them, though movies and the music halls and vaudeville attracted many patrons. Students of musical and theatrical inclinations also staged their own productions. One of the more ambitious, created and performed at the University of Toulouse, was the musical comedy, "Getting Toulouse," described as "a delicately shaded portrayal of the more salubrious phases of A.E.F. life." The words and music were written entirely by the men at the school, some with experience working on Broadway. One of these was Sergeant Clark E. Biggs, who wrote the book, and was also the show's producer. He had had a hand in producing some of the big musical comedy successes in America before entering the army. The cast numbered thirty-five men, playing both male and female parts, and "it was discovered that many of these girls had been hiding their true charm under the camouflage of a moustache, but visions of Paris have induced them to return to their virginal beauty." The costumes were designed by members of the School of Beaux Arts, and made in Toulouse at the cost of several thousand francs.[78] Not having a large student population to draw from, the students at the University of Poitiers contented themselves with putting on the much less demanding play, "The Man From Home," in late June at the City Theater in Poitiers, for the benefit of the townspeople and the student body.[79]

And there were numerous glee clubs. The one at Poitiers was repeatedly featured at the local cathedral, while the thirty-man club of the University of Grenoble made a tour of all the soldier detachments in France.

The "Whizzbang Jazzers" at the University of Toulouse were typical of another sort of musical organization established at many of the universities where the Americans were. Indeed, jazz was all the rage by this time, both in America and, increasingly, in Europe.[80]

In addition, numerous bands, orchestras, and choruses were organized, which performed at various functions. The students often took opportunities to involve the local French population in the staging of social affairs, which more often than not featured musical performances and dances. At Poitiers, for example, much effort was expended in staging a series of elaborate balls, marking the return to a social season, and a state of "normalcy." The balls were "for all of our French friends who for over four trying years have watched with fear and pain the rise and fall of the struggle of nations," and would therefore be events that "will be greeted with pleasure, for it will recall the days before the war when all France was a land of sunshine and laughter." As for the doughboys, "who have been drifting, drifting, for a longer time than we care to remember, down the sluggish current of military affairs," the events would bring relief as well. The premier event came off splendidly, and the jazz band especially delighted the French, as did other musical groups that performed, and the subsequent series of balls was a great success.[81]

One of the major aspects of the American involvement in England was extensive travel. Though not to the same extent, lacking the unlimited travel passes issued in Great Britain, and involved in more closely focused courses of study, the students in France nevertheless made much of travel opportunities. At Montpellier, for example, two-day trips were scheduled for Nîmes and Pont du Gard—styled "Roman excursions"—and then to Carcassonne, with its medieval setting.[82] At Poitiers, there were frequent shorter trips to the castle at Touffou, to Bonnes in the valley of the Vienne River, and to the historic city of Châtellerault.[83] At Toulouse, Lourdes was a favorite travel destination, which many visited during the Easter week. More for recreation than for study, students spent their weekends in numerous places, such as Pau and Biarritz. So popular was Biarritz, that the Knights of Columbus began running free excursions from the city to the Spanish border.[84]

One of the more ambitious trips, for both educational and vacation purposes, was arranged during the Easter vacation for the student detachments at the sister universities of Aix and Marseilles. About 150 officers and men proceeded by rail to Toulon where they boarded two French naval vessels, the *Bellatrix* and the *Opiniâtre*. The two torpedo boats sailed to the island of Corsica, landing at Ajaccio where four days were spent in educational tours. The students then departed on board the two French ships for Nice, and back to Marseilles by train. The trip was apparently an exceptional one; the men were well received everywhere, and receptions and other entertainments were held for them. The students who did not take the cruise spent Easter vacation on an educational tour to the cities of Arles and Nîmes.[85]

The students saw not only something of France, but also much of the French. The majority of the students were housed with French families, forging many life-long friendships. This close connection with the French also contributed to good behavior on the part of the Americans. Always mindful of Franco-American relations, GHQ could only applaud these developments. Lieutenant Colonel Gordon R. Catts, making an inspection for G-5, asserted that "the association of the students with the best families . . . is a great factor in discouraging drinking and other excesses which might be expected [otherwise]."[86]

The Americans met the French at every level. They scheduled chess matches with them and were instructed in fencing by French experts. They in turn entertained the French with dances and other social events. Their glee clubs sang for them and their bands and musical ensembles serenaded the French, informally and formally. They worshiped with them. They had a close contact with French students, often forming clubs which had both French and American students as members, and they published student papers together. Indeed, one French professor noted that he had never seen French students working harder, "owing solely to the fact that their efforts were stimulated by the example set them by the American students."[87]

The French responded similarly, hosting many receptions and in other ways recognizing the American presence among them. And there were several marriages between French girls and American doughboy students, though apparently these were not as numerous as in England. Altogether, the close contacts with the French aided considerably in the attainment of the American venture's fundamental aims: that the doughboys should encounter French life, language, and culture in significant ways.

To be sure, there were some conflicts between American soldier-students and the French, though none apparently too serious. And some of the customs seemed strange, that of kissing on the cheeks, for example, which one anonymous poet could only lament:

A La Bouche

If a body meet a body
    Coming down the rue
If a body kiss a body,
    That's a "How-de-do";
But kissing à la Française
    Isn't kissing true;
One kiss upon the bouche is worth
    The two upon the joue.

. . . .

As he concluded: "For kissing à la Française/I surely can but see/Is only half the fun of old/Not what's smacked up to be."[88]

The French had to make allowances as well, as the editor of *Qu'est-ce Que C'est?* acknowledged. Observing that the French had "opened not only their doors, but their hearts to us," they also had to make "full allowance for our lack of knowledge of their customs, and our partial ignorance of their language." When the Americans returned home, they would always remember the "charming city" of Toulouse, "where the spirit and the grace and the culture of the foremost country amongst civilized nations was expressed to us by the man in the street and the master in the study alike." In fact, as one wag observed, among the "Famous Friendships" in the history of the world, such as Damon and Pythias, David and Jonathan, Barnum and Bailey, not to mention "rank and file," there must now be added: "*étudiants américains et petites Toulousaines.*"[89]

To the editor of *Qu'est-ce Que C'est?* the French had taught some fundamental truths, such as "that there is something in life besides the eternal rush of the western world, the mighty seeking of the mighty dollar. You have shown us that it is possible to enjoy life on little material wealth—you have taught us some of the joys of the leisurely life." Furthermore, Americans had come to realize that art was "not the strange, mystical thing which some 'back home' would have it; and that it is not unmanly to appreciate, or be a worker in, the finer fields of life." In fact, Americans were surprised to learn, "some of the manliest men we have met among you, are artists." The same was true of opera performers who were hardly "effeminate because [they] sing, instead of ploughing land or building houses." And Americans had learned something about relations with women. In France, they were surprised to discover that French women were partners with their men; "they share in and aid in your successes, and we've seen how bravely they bore with you in days of sorrow." Many American doughboys were therefore better instructed, "so, when we go home we're going to be more friendly with our women, we're going to make 'pals' of them, not merely prostrate ourselves at their feet, as some Tennysonian knight of old would have done."[90]

As a token of their appreciation, the American students at the fourteen French universities subscribed fourteen 1,000-franc scholarships to be awarded to one student at each of the universities, either a soldier or a son of a soldier of the war, to attend any American university of his choice. In addition, an organization was to be formed in the United States to encourage and aid the exchange of students between the two countries in the future. The scholarships were in lieu of bronze plaques that were to have been presented to each of the schools. It had proved impossible to obtain these in time for formal presentations as originally planned by GHQ.[91]

There is little doubt that the program of permitting American students to study in France triumphed. Colonel Charles W. Exton, of G-5, GHQ, was of this opinion, and he also affirmed that, in his view, the Poitiers detachment was the best of the fourteen detachments in France. It seemed to him to exemplify best what the experience had meant, both for the American soldier-students and for the French involved. He was especially pleased "with the spirit with which the Yanks had entered into the French university life, and with the splendid relations established between the French civilians of Poitiers and Americans." Miss Nan Cannon, a former Ohio newspaperwoman, then with the publicity department of the American Educational Commission, was similarly impressed with the Poitiers detachment, stating that the work there "stands out among the foremost in France."[92]

However, Lieutenant Colonel Catts, who inspected several of the student units—though apparently not the one at Poitiers—had his own opinions, asserting that the conditions at Montpellier were better "in every way" than at any others that he saw. In addition, "the relations with the French students and civilian population could scarcely be better," he stated.

In fact, Catts was generally favorably impressed with the several American student detachments that he inspected. Though there had been some initial difficulties in placing students with families and in generally getting the men organized, these were soon resolved, and there was subsequently little trouble at any of the universities. Students were, for the most part, enthusiastic about their opportunities to study at the venerable French universities, considering themselves "extremely fortunate." They justified their selection by exemplary conduct and hard work. As a rule, the relations with the French were likewise cordial, and lasting beneficial results were anticipated. In some of the cities near leave areas, for example, where "unfortunate breaches of discipline which might have prejudiced the French have occurred," there had been some hesitation by the French in establishing close contact with American troops. However, it was soon apparent that the students "were representative of our best young manhood," he observed, and accordingly the French responded with cordiality and even enthusiasm to their presence. Catts concluded, therefore, that this venture in higher education was a resounding success.[93]

The fact remains, that the vast majority of the rank and file of the AEF did not meet life in Europe on the same level as did the college and university men. But those few thousand collegians, while amounting to a minuscule percentage of the two million men who made up the AEF, were among the brightest of the force, and the results of their encounters with European higher education were of lasting significance. They were forced to examine their lives from a higher plane. To many of them, this was certainly a revelation and a positive benefit for their futures and that of the nation. Their involvement in European intellectual and social life is a testimony to the wisdom and initiative of concerned leaders both within and outside of the army, to force the war to pay off in more than the dismal casualty figures and the many other corrosive effects of such a protracted and profound conflict.

NOTES

1. MacLean was under Erskine's control, but the latter does not seem to have been closely involved with the operations in Great Britain. See account in *Qu'est-ce Que C'est?* (student paper of the American detachment at the University of Toulouse), vol. 1, no. 4, April 9, 1919. The American University Union was founded in Paris in July 1917, shortly after America's formal entry into World War I. It grew rapidly, enrolling thousands of college and university men from about 150 colleges and universities as members in the Paris club that it became, maintaining club rooms and a headquarters in the French capital. Its Paris head was George H. Nettleton, former professor of English in the Sheffield Scientific School, Yale University.

2. *The American Soldier-Student*, no. 6, June 18, 1919. This was a weekly newspaper, published every Wednesday in London to serve the entire student detachment of the U.S. Army in Great Britain. Some seven issues were published, the last—no. 7, dated June 25, 1919—was a twelve-page paper, containing much information on the short but lively history of the detachment.

3. This information is ibid., no. 7, June 25, 1919.
4. Ibid., nos. 3, and 7, May 28 and June 25, 1919.
5. Ibid., no. 4, May 28, 1919.
6. Ibid., no. 7, June 25, 1919.
7. Article summing up their British experience, ibid., no. 6, June 18, 1919.
8. Ibid., no. 3, June 4, 1919.
9. Ibid., no. 7, June 25, 1919.
10. Ibid., no. 6, June 18, 1919.
11. Ibid., no. 5, June 11, 1919.
12. Ibid.
13. See account ibid., no. 3, May 28, 1919.
14. Account ibid., no. 5, June 11, 1919.
15. Article ibid., no. 4, June 4, 1919.
16. See several accounts ibid., nos. 3, 4, 6, and 7, May 28, and June 4, 18 and 25, 1919.
17. Articles ibid., nos. 5 and 7, June 11 and 25, 1919. Among other things, the English-Speaking Union proposed that members jointly celebrate such national holidays as Washington's birthday, Shakespeare's birthday, Magna Charta Day, Empire Day, American Independence Day, and American Thanksgiving Day.
18. See articles ibid., no. 6, June 18, 1919.
19. Article, ibid., no. 7, June 25, 1919.
20. Articles and editorial ibid., no. 4, June 4, 1919.
21. See account ibid.
22. Ibid., no. 3, May 28, 1919.
23. Ibid.
24. Ibid., nos. 4 and 7, June 4 and 25, 1919. There is a detailed analysis in issue no. 7.
25. Ibid., no. 7, June 25, 1919.
26. Ibid., no. 6, June 18, 1919.
27. Ibid.
28. Ibid.
29. Ibid.
30. Ibid., no. 4, June 4, 1919.

OTHER UNIVERSITIES

31. Ibid., nos. 5 and 7, June 11 and 25, 1919.

32. Ibid., no. 5, June 11, 1919.

33. Ibid., no. 6, June 18, 1919.

34. Ibid.

35. Article ibid., no. 7, June 25, 1919.

36. Ibid.

37. Ibid., no. 5, June 11, 1919. Only one member of the student detachment in Britain died during the stay: one Corporal Sharrar, who died shortly after his arrival at Knotty Ash Camp of pneumonia. See article ibid., no. 7, June 25, 1919.

38. See short historical account in *Qu'est-ce Que C'est?* (student paper of the University of Toulouse American students), vol. I, no. 4, April 9, 1919.

39. See pamphlet, *University Work in Paris For American Soldiers*, published by the Y.M.C.A. Army Educational Commission, A.E.F., in folder "College of Fine Arts, etc.," Box 1924, Entry 408; and *Stars and Stripes*, May 2, 1919, for article regarding the one thousand students studying at Paris. Most were attending the Sorbonne and were living in the Latin Quarter, around the boulevards of St. Germain and St. Michel. Some were in the homes of French people; some were in hotels in the Latin Quarter. Some of the students planned to demobilize in France and continue their studies there. Professor S. H. Bush was dean of the American detachment at the Sorbonne, and Major J. L. Coolidge was the educational commandant of Paris. Colonel Charles W. Exton, as G-5, GHQ, representative in Paris, was the military man in direct charge of all student activities in the A.E.F., including the supervision of all the American school detachments at the several French universities, though Erskine was in charge of the educational side of things.

40. See memorandum, Lt. Col. Gordon R. Catts, G-5, GHQ, AEF, to John Erskine, educational director, American E.F. University, May 6, 1919, in folder "Reports," Box 1965, Entry 420.

41. See account in *Les Beaux Jours*, a paper published weekly by the French and American students at the University of Poitiers, vol. 1, no. 4, April 19, 1919. The number of advanced students at Poitiers was 32 out of the 249 soldiers enrolled there.

42. See letter, Erskine to Frederick P. Keppel, third assistant secretary of war, March 19, 1919, in folder "Publicity and Public Press," Box 1901, Entry 408; and Memoranda, John Erskine, educational director, American E.F. University, to General Rees, May 3, 1919; and Lt. Col. Gordon R. Catts, G-5, GHQ, AEF, to John Erskine, May 6, 1919; and lengthy report Catts to assistant chief of staff, G-5, of an inspection trip to the university detachments at Aix-Marseilles, Montpellier, Grenoble, Lyon, and Besançon, May 5, 1919, in folder "Reports," Box 1965, Entry 420.

43. *Les Beaux Jours*, vol. 1, no. 3, April 12, 1919.

44. Article in *Qu'est-ce Que C'est?* vol. 1, no. 3, April 2, 1919. At Toulouse, the first commandant was Major John De Hart Harrison, followed by Major J. H. Wallace.

45. Report, Catts, May 5, 1919.

46. Ibid.

47. General Orders No. 69, GHQ, AEF, April 21, 1919, *United States Army in the World War*, vol. 16, pp. 737-38.

48. Report, Catts, May 5, 1919.

49. *Les Beaux Jours*, vol. 1, no. 5, April 26, 1919.

50. *Qu'est-ce Que C'est?* vol. 1, no. 3, April 2, 1919. Professor K. C. Cowdery had been dean at Toulouse earlier but had resigned to go to Beaune as professor of Romance languages. Dawson, who had been given a six-month leave by Howard College in order to serve as a dean in the AEF, had, for a short time been at the

University of Bordeaux.

51. *The Soldier-Student*, vol. 1, no. 3, April 5, 1919.

52. *Qu'est-ce Que C'est?* vol. 1, no. 6, April 23, 1919.

53. See editorial ibid., vol. 1, no. 3, April 2, 1919. Parmer was the editorial editor of the paper.

54. Ibid., vol. 1, no. 5, April 16, 1919.

55. Ibid., vol. 1, no. 7, April 30, 1919. This editorial was by Pvt. Donald M. Calley, the paper's news editor. It received a ready response, many readers telling the editor that it was the best thing the paper had published. See ibid., vol. 1, no. 8, May 7, 1919.

56. Editorial, "We've Learned a Little, 'Over Here,'" ibid., vol.1, no. 8, May 7, 1919.

57. Article ibid., vol. 1, no. 9, May 14, 1919.

58. See, for example, editorial ibid., vol. 1, no. 11, May 28, 1919.

59. Editorial, "We Are Americans!" ibid., vol. 1, no. 6, April 23, 1919.

60. Fred Davis Baldwin, "The American Enlisted Man in World War I," Ph.D. dissertation, Princeton University, 1964, p. 240.

61. Ibid., p. 216.

62. Ibid., pp. 233-34.

63. *Qu'est-ce Que C'est?* vol. 1, no. 8, May 7, 1919.

64. See editorial, "The A.E.F. and the Church," by Private Donald M. Calley, ibid., vol. 1, no. 10, May 21, 1919.

65. See discussion ibid., vol. 1, no. 3, April 2, 1919. In addition, Americans had learned something of morals. Unfortunately, some of the "would-be moralists at home still believe that it is a crime to enjoy life in a pleasant way," one American wrote. "You've shown us differently," he continued, "and we have a feeling that the Almighty God admits as many of you through the Heavenly Gates as he does of us—even if you do make of Sunday something more than a solemn day of prayer."

66. *Les Beaux Jours*, vol. 1, nos. 4, 8, and 11, April 19, May 17 and June 9, 1919.

67. *Qu'est-ce Que C'est?* vol. 1, no. 10, May 21, 1919.

68. Ibid., vol. 1, no. 4, April 9, 1919; ibid., vol. 1, no. 10, May 21, 1919.

69. *Les Beaux Jours*, vol. 1, no. 5, April 26, 1919.

70. *Qu'est-ce Que C'est?* vol. 1, nos. 3 and 5, April 2 and 16, 1919; *The Soldier-Student*, vol. 1, no. 3, April 5, 1919.

71. *Qu'est-ce Que C'est?* vol. 1, nos. 4 and 10, April 9 and May 21,1919.

72. *The Soldier-Student*, vol. 1, no. 3, April 5, 1919.

73. *Qu'est-ce Que C'est?* vol. 1, no. 12, June 4, 1919.

74. *Les Beaux Jours*, vol. 1, no. 7, May 10, 1919.

75. Ibid., vol. 1, no. 6, May 3, 1919.

76. See article in *Qu'est-ce Que C'est?* vol. 1, no. 3, April 2, 1919.

77. *Stars and Stripes*, May 9, 1919.

78. *Qu'est-ce Que C'est?* vol. 1, nos. 5, 7, and 10, April 16, April 30, and May 21, 1919.

79. *Les Beaux Jours*, vol. 1, no. 11, June 9, 1919.

80. *Qu'est-ce Que C'est?* vol. 1, no. 8, May 7, 1919.

81. *Les Beaux Jours*, vol.1, nos. 4, 8, and 9, April 19, and May 17 and 24, 1919.

82. *Soldier-Student*, vol. 1, no. 3, April 5, 1919.

83. *Les Beaux Jours*, vol. 1, nos. 3 and 4, April 12 and 19, 1919.

OTHER UNIVERSITIES

84. *Qu'est-ce Que C'est?* vol. 1, no. 11, May 28, 1919.
85. Report, Catts, May 5, 1919.
86. Ibid.
87. Ibid.
88. *Qu'est-ce Que C'est?* vol. 1, no. 3, April 2, 1919.
89. See ibid., vol. 1, no. 3, April 2, 1919.
90. Ibid., vol. 1, no. 13, June 11, 1919.
91. Ibid., vol. 1, no. 12, June 4, 1919.
92. *Les Beaux Jours*, vol. 1, no. 11, June 9, 1919.
93. Report, Catts, May 5, 1919.

CHAPTER 9:
Last Days of the AEF University

LATE IN THE TERM, the students at Beaune were canvassed as to whether or not a second session would be desirable.[1] However, it was apparent that the need for a full-scale university in the AEF was rapidly diminishing, and a second term was unlikely. Already, it was proving difficult to maintain stability among students and staff. Replacements for the faculty were especially hard to obtain. Therefore, in early April, applications of instructors to return to the United States were no longer approved, barring exceptional circumstances. This policy applied to both officers and enlisted men even if their organizations had already sailed or would do so before June 7. Applications from students continued to be accepted, however, in cases of distress or emergency, or if their respective units were returning home.[2] In fact, as he had promised in the beginning, Reeves was determined that no student would be deprived of the privilege of returning to the U.S. with his organization should it leave prior to the term's end.[3]

By mid-April, the roster of the university community reached a maximum of 15,895 people; by early May it had sunk to 12,153, and by the end of the month, 11,601.[4] The mid-term reports recorded that there were 504 instructors at the university—excluding Allerey—including officers, enlisted men, and Army Educational Corps personnel, with eighteen student instructors supplementing the number. Administrators, clerical staff, and others brought the total staff to 683 people. Of this number, 299 were officers, 225 were enlisted men, and 127 were members of the Army Educational Corps, while 32 were French citizens. It ranked in size with the largest universities in the United States.[5] By this time, the university—excluding Allerey—had 13,108 student registrations— not registrants, as each registrant was taking from one to four courses—enrolled in 397 classes, of 220 courses, taught by forty-one departments in eleven colleges: agriculture, art, business, education, engineering, journalism, law, letters, medicine, music, and science. Of the 13,108, satisfactory grades were turned in at mid-term for 11,222, while 889 were unsatisfactory and 997 were absent or had no grades recorded. These figures did not include the post and division schools, the Farm School at Allerey, or the citizenship program.[6]

By the end of April, the fate of the university was clear: "unless there are unforeseen military contingencies," it was to close "at the end of the present term." This resulted mainly from the paramount fact that the school was "never intended to delay for one moment the return of any soldiers to the United States."[7] All personnel connected with it, whose organizations had already departed for home, were to be organized into provisional regiments for return to the U.S. as a body. The separate units would be made up of officers and men from individual states, with larger organizations, such as battalions, grouped according to districts, for example the one for New England.[8] Reeves promised that time immediately after the school's closing would be used for a termination of its total affairs, "but this time will be reduced to the minimum."[9]

But instead of a general slackening of focus, on the contrary, all concerned were ordered to expend every effort to utilize all the facilities of the school to the fullest extent possible. The classes would continue, "with renewed energy." No action was taken that would in any way lessen the efficiency or the continuation of the educational work. No new building projects were authorized unless absolutely necessary for classes, though the construction work underway would be completed. The work pertaining to the planting of gardens and beautifying the grounds would be continued, however, as originally planned.[10] As to the students, attempts would be made to bring in more of them, especially from the combat divisions and the Services of Supply [SOS] for short-term courses.[11] Some of the music courses, earlier canceled because of a shortage of instruments, such as violins, were resumed in mid-May when the College of Music found a few additional instruments for student use.[12] Students then began receiving individual instruction.[13]

Orders required that all equipment be kept in good repair and properly painted, oiled and maintained, and stenciled or hand-lettered with appropriate college markings. Discipline had to be strictly enforced, and a high state of military appearance and bearing was to be maintained, "up to and including the hour of discharge from the service." The specified time allotted for military instruction would be adhered to. Timely requisitions would be made for properly uniforming all enlisted men, in the "best uniforms obtainable in the A.E.F.," so that they might present "the best possible appearance at all times." Naturally, the officers would continue to set a proper example in all respects.[14]

Still, as a sign of the times, all directors of colleges were instructed to make estimates of their requirements of crates and boxes for packing their equipment, and to list all equipment, machinery, textbooks, fixtures, and furniture not required for the proper functioning of their colleges for the remainder of the term.[15]

In the event that some of the men desired to remain in the army, a recruiting office was opened on the campus. Fit men ages 18-40, with no dependents, could enlist for one or three years, in the regular army, for service in the Army of Occupation in Germany, or for other assignments.[16] Also, the men were informed that for others who might desire to remain in Europe, fellowships were being offered to a few men in the army. Soon, 1st Lieutenant Fred Gasser, Medical Corps, was dispatched to Beaune as a recruiting officer, and Reeves was informed that "it is desired that every aid be given to enable him to conduct an energetic campaign throughout your command."[17]

New arrangements were made to meet religious needs. Large, united congregations were subsequently assembled in the University Theater for services, some featuring music by the school's band and orchestra. In addition, Bible classes and Sunday schools met in regimental huts, and Catholic masses were scheduled.[18] The first of the large meetings occurred on May 18. The first sermon was preached by Chaplain F. K. Little, of the 10th Regiment. He considered the future of the world and the effect of the return of the men of the overseas forces and their spiritual attitudes, which he saw as a positive force on American life.[19]

As the time for closing drew near, the pace accelerated. Officers living in Beaune were instructed to obtain receipted bills from their landlords prior to the time of their departure, with clear indication that no claims for damages to property were outstanding. All business relations with merchants were to be closed with nothing remaining for adjustment.[20]

No more leaves were granted except for engineer troops and hospital and permanent personnel. Even passes were limited.[21]

In order to begin the reduction of personnel, when the numbers of students began to decline drastically by mid-May, a substantial reduction in faculty was made. Officer and enlisted instructors were thereafter permitted to return to their own organizations as soon as their services could be spared.[22]

The Farm School at Allerey was closed on May 31, and all personnel transferred to Beaune. Naturally, every effort was made to leave the buildings and grounds at Allerey "in the best possible condition."[23]

In anticipation of the forming of the casual companies, instructions were published specifying the authorized and necessary equipment for each man as provided for by embarkation regulations. Prior to departure for ports of embarkation, the troops would be inspected "with a view of eliminating unauthorized government property and excess personal property from the baggage or freight of the returning men."[24] Only personal effects that could be carried in packs were permitted, though a special dispensation was later obtained allowing enlisted students at Beaune and Allerey to take with them the paintings, mechanical drawings, and small models that they had produced as well as personal textbooks.[25]

Library officials soon took steps to collect books that were due.[26] It was necessary to request that all library books be returned by 1 June; subsequent checkouts would be permitted only on special request and then only for periods of twenty-four hours. The library, however, would remain open "as long as justified by the use made of it, for reference and general reading." It also furnished each company leaving for ports of embarkation with sufficient magazines to read during the journey.[27] To reduce its considerable holdings, the library made a present of a thousand books to the city of Beaune at the Memorial Day ceremonies.[28]

Soon other offices were limiting services or closing. As of June 1, the dental clinic accepted only emergency cases or those that could be completed in one sitting.[29] Textbooks were turned in by June 4 to the issuing offices.[30] All post exchanges closed as of May 30, followed by the sales commissary on June 9.[31] The last classes were held on May 29, the final exams being scheduled for the class periods of June 2 and 3. All athletic supplies were due in on June 9, except for the equipment that would be used on the way to the embarkation ports.[32]

Sunday, June 1, was the day set aside for the reorganization of the student body into casual companies of 150 men each. No passes were issued and no person was allowed off base on that date.[33] University officials were assisted by three officers from St. Aignan's First Replacement Depot who spent the last few weeks at Beaune preparing the reorganization into companies.

Embarkation instructions also required that everyone was to be physically examined and deloused, with a certificate being issued to that effect. All personnel, without exception, also had to be vaccinated for typhoid or para-typhoid, with a certificate issued for this also. Thus, a

169

"shot in the arm" was a necessary preliminary to the return home.[34]

Soldiers with French wives could take them with them, and a special bridegrooms' casual company was formed. Wives and their husbands were scheduled to sail home on the same ship.[35]

Other special organizations included the Marines, numbering about 75 men. These were organized into their own company and housed in separate billets. Their destination in the U.S. was Quantico, Virginia.[36]

The university's life ended in a flurry of intense activity. For one thing, the school held a large track meet on June 4 and 5, which was enthusiastically attended. Medals, serving as something "in the nature of a souvenir of the University," some of which were contributed by the French Homes Committee, were awarded to the winners. The citizens of Beaune were invited, and the university band furnished stirring music for the occasion.[37]

More important were the Inter-Allied Games held in Paris May 21-June 1. All members of the university were eligible to try out for the meet. A large contingent industriously trained for the events, some of the men eventually competing. They placed only third in the medley relay, however, and won no other medals. The results were disappointing because several of the school's runners had placed well in preliminary heats.[38]

Another late event, beginning on June 4, was a four-day Inter-Allied Conference on World Agriculture. Major speakers from the U.S., Canada, Britain, and France addressed the world's agricultural problems in the emerging era of peace.[39] One of the featured speakers was Herbert Hoover, who delivered the closing address.[40]

This was one of the last of a series of conferences at the school. It had previously hosted a number of meetings concerned with problems facing the nation and the world, mirroring in this way as well its academic contemporaries in the United States. In mid-April, a conference of army school officers and Educational Commission advisors from throughout the AEF discussed the problems and progress of army schools. One problem addressed was how army and other American schools should confront a question plaguing educators, i.e., "Who won the war?" The conference concluded that the chauvinism, which was much in evidence regarding the matter, must be considered inappropriate in American schools. Indeed, this "national flag-waving of the junker [sic], 'Uber Alles' type, as the underlying basis for the teaching of history," was strongly deplored. As to the central question, who had won the war, the provincial attitude of writers intent upon glorifying the achievements of their own country at the expense of others, led to fears that nations had learned little from the war. The result would be that the victorious states would all claim to have defeated the Central Powers. This would produce "at least, a hot argument, and at the most, a breeding of suspicion and narrow-minded nationalism with the ever-present danger of engendering a new rumpus." Germany's attitude before and during the conflict was the best example of what must be avoided. Therefore, the roles of all the Allied nations in the struggle must be clearly described, and each given credit for its contributions. The conference urged all of the Allied nations to adopt this sane stance.

The conference also considered the future of American education under the impact of the war experience. One of the principal speakers addressing the subject was Professor John Erskine. He was gratified that "the Yank, it is found, is proving as apt a pupil in education as he was in warfare." The experience of the army schools had revealed, he continued, that American soldiers know what they want, and therefore studies "should be provided in America to take care of those wants, rather than. . .arbitrarily [hand] out courses." He suggested, for one thing, the revamping of modern language study, and, reflecting the principles of admission and the conduct of courses at the AEF university, he urged that all who desired it, "should be given [the] opportunity to study what [they want] at any time and with no age limits." Continuing adult education, without regard to the usual admission standards, was the ideal that he extolled.

Elaborating upon the idea of the army as the "khaki university," other discussants advanced the notion that the army had an obligation to each soldier to make of him a practical electrician, a mechanic, or an expert in some other line that he might elect, and to train him so that he could get a good job when he left the army. Colonel Reeves further insisted upon the training of officers in business skills, because "the war . . . has shown that officers are needed for other than combat divisions." Future military leaders must be able to perform the increasingly complex duties of the SOS or its successor, he declared.

To help solve the world's pressing educational problems, the conference adopted a memorandum urging the establishment of a permanent bureau of education within the League of Nations, which would convene frequent international educational conferences. This was formally presented to the Paris Peace Conference by the Army Educational Commission.[41]

Another agrarian event in these days was an agricultural exposition set up in one of the large buildings on campus. The exhibits were created by personnel from the Farm School at Allerey and included a miniature model American farm and a life-sized kitchen with modern equipment, as well as other rooms, and buildings, with additional exhibits, well illustrated by photos, of fruit growing, bee keeping, poultry raising, soil fertility, and forestry.[42]

In this connection, the College of Agriculture, aided by women from the YMCA and the Army Educational Corps, also staged a "mild rural comedy," *Back to the Farm*, a "propaganda play" that advocated the installation of modern farming methods. This "charming comedy of rural life," was written by Merline H. Shumway while enrolled at the Minnesota Agriculture College, and was first produced by the students of that school. The stage sets at Beaune were painted by German prisoners-of-war who were professional painters and camouflage artists, and who were then "on detached service" at the university. The play was well received, and its message was presented "without making it tiresome." Before the show, the Barn Yard Quartette sang some numbers.[43]

Also, to the very end, the university sent out teams to hold farmers' institutes. These were especially active in the 3rd Army.[44] Other speakers were sent throughout the AEF to lecture on such subjects as labor conditions, public health and welfare, and commercial and business opportunities. These too conducted two- or three-day institutes, many of which appeared under the auspices of the Citizenship Department of the Army Educational Commission, headed by Dr. John A. Kingsbury.[45]

Last-minute educational tours continued to be scheduled. For example, fifty students of the animal husbandry class of the College of Agriculture visited a large showing of Percheron stallions conducted by two of the largest exporters of the breed in France at Le Ferté Bernard. Journalism students journeyed to Besançon to visit the historic city, while twenty-eight students and instructors of the physics department visited the Schneider munitions plant, the largest in the world, located at Le Creusot.

Numerous social events and other entertainment crowded the activities calendar. Miss Elsie Reed of the Coast-to-Coast Company of players and performers entertained in various places on campus, headlining a variety show. The "Doughboy Follies," a traveling doughboy show, featuring vaudeville, minstrelsy [sic] and jazz, performed at the University Theater before an S.R.O. crowd on Saturday evening, May 10. Also appearing about this time was a quartet of entertainers, billed as the Home Folks, who performed "light Chautauqua" rather than the more customary vaudeville. The university band opened a series of Monday evening concerts at the University Theater on May 12. The Knights of Columbus sponsored two proms, one for the officers and another for the enlisted men. The Knights also maintained an officers' club in Beaune, which featured weekly dances. The YMCA managed five recreation huts staffed by American girls in the regimental area. These held fudge parties, afternoon teas and evening dances.[46]

In downtown Beaune, the Y operated two large club houses, which were particularly active in the last days of the university's life. Both on the rue de Gigny, these had been opened in the latter part of April. The Enlisted Men's Club was located in an old chapel and parish house, surrounded by lilacs, shrubbery, and "inviting shade trees," which had been especially renovated for the purpose. The club featured dances and dinners every evening, as well as luncheons on Saturdays and Sundays, two hundred men being accommodated at every meal. Just across the street was the Officers' Club located at Villa Phillipe, an imposing old mansion surrounded by a shaded park.[47] The French-American Club, where American students could study French, was located on the rue Paradis, in a fifteenth-century Burgundian palace. The club scheduled dances and other events calculated to improve Franco-American relations.[48] Also active were the club rooms of the Jewish Welfare Board, located in the rue Mafoux in Beaune. Here meals were served, dances held, and other entertainment provided.[49] Jewish soldiers also organized themselves into the Menorah Group for the study and discussion of contemporary Jewish problems.[50]

Perhaps most memorable of the events in the last days were those involving the French. On May 22, the university hosted "Beaune Day," at which time many citizens of the town inspected its buildings and grounds. The event was sponsored by the French Relations Committee, chaired by George S. Hellman, director of the College of Fine and Applied Arts. An American soldier fluent in French was on duty at each of the classrooms and barracks to act as interpreter and guide. The day's events closed with a band concert at the University Theater, followed by refreshments and a parade by the 9th Regiment on Pershing Field.[51]

On May 29, a group of distinguished French educators descended upon the campus for an inspection tour, review, dinner, and reception. The delegation was headed by M. Lafferre, the minister of public instruction for France. Others of note included Du Courtois, *chef adjudant du cabinet du ministre des travaux publics*, who represented M. Claveille, minister of public works; Louis Aubert, who represented M. Tardieu, the high commissioner for Franco-American affairs; and prominent local officials. At a review on Pershing Field, Lafferre decorated twenty-three officers and members of the Educational Corps for distinguished services in academic work. In the name of the president of the French Republic, he bestowed the order of the *chevalier de la Légion d'honneur* upon General Robert I. Rees, the superintendent of education in the AEF; Colonel Ira L. Reeves; and Dr. John Erskine. The university's executive officer, Major Watrous, Dr. Kenyon Butterfield, and Dr. Frederick Ellsworth Spaulding all received the decoration *officier de l'instruction publique*. The other university leaders, seventeen in all, were awarded the distinction *officier d'académie*. The latter two decorations had been created by Napoleon at the same time that he created the *Légion d'honneur*, his intention being to recognize the merits of outstanding French educators and writers. After the ceremonies, a formal dinner was held at the Hôtel de la Poste in Beaune.[52]

A few days later, on June 5, Mayor M. J. Vincent, members of the city council, and prominent citizens held a formal reception honoring the Americans at the city hall. It too was followed by a banquet at the Hôtel de la Poste. A resolution inscribed on sheepskin, and framed with manifestations of grapevines and grapes symbolizing Beaune's main industry, was presented to Colonel Reeves. Its text, in French and English, referred to the common war efforts of Frenchmen and Americans. Indeed, American doughboys, whose courage and bravery had "largely contributed to the triumph of right and civilization over barbarity," would long be remembered, as would the university, which had "astonished everybody." In addition, Reeves and Erskine, as well as General Rees, were made honorary citizens of the city.[53]

Reeves now departed, and Major Andrew J. Bush of the Quartermasters, took charge. He commanded a Quartermaster unit, a Motor Transport Corps, and other support detachments numbering just over 1,000 officers and men. The district engineer detailed three officers and six men to Bush's command, and the university surgeon ordered two officers and eight men to remain on duty. All other auxiliary troops, numbering about 77 officers and 1,361 men, were relieved from further duty at Beaune, and proceeded to Le Mans for reassignment.[54]

The ending of the university forced the disposal of the structures and equipment at both Beaune and Allerey. The final arrangements were placed in the hands of Major Richard Brooke, the district engineer officer at Beaune, who worked with the French regional commander at Dijon.

LAST DAYS

The French purchased many of the buildings to be dismantled and reassembled in the devastated areas of France, where they would house refugees until permanent housing could be built. Thus, "buildings where University students danced," reverted to other, less festive, uses. Certain other supplies were transferred to the French army, though some equipment was sent to the American Army of Occupation in Germany.[55]

But none of the last-minute flurry of activities could measure up to the sound of the words "Going Home." As the university's newspaper joyously stated, "the hills echo the words; the atmosphere is surcharged with happiness that the soon-to-be-discharged doughboy can scarce contain."

Though there might be slight tinges of regret, the personnel were ready and eager to begin the journey home. The reforming of the student body into casual companies was done with alacrity and the full cooperation of all concerned. Indeed, "never before has the doughboy been so thoroughly in accord with the re-formation of his 'outfit' as he evinces in his readiness to forget his affiliation with his 'Provisional' and become part of a St. Aignan casual company."[56] It was stipulated that all Army Educational Corps people were likewise to go home at this time, except those desiring discharges in France. Those concerned subsequently reported to the university for discharge before June 10. They were scheduled for shipment home, "on or about June 15."[57]

The university officially ceased to exist at 6:00 P.M., Saturday, June 7, 1919.[58] Remaining personnel reverted to a more traditional military command, and the evacuation of the campus began on the following day. All that remained were the memories, and the record of "the mental development and improvement of its thousands of students." Afterwards, the university would persist as "a monument of the most lasting nature," one writer observed, whose effects would follow to every section of the U.S. But for the moment, all of these considerations were dwarfed by the realization that, "with ordinary good fortune, most of us will celebrate July 4 on American soil."[59]

SOLDIER-SCHOLARS

NOTES

1. Memorandum, from General Rees to Colonel Reeves, GHQ, AEF, April 25, 1919, in folder "A.E.F. University. Historical Data," Box 1956, Entry 409.
2. Memorandum No. 23, Headquarters, American E.F. University, April 2, 1919, Entry 415.
3. Bulletin No. 76, Headquarters, American E.F. University, April 30, 1919, Entry 414.
4. See "History of Medical Department, American E.F. University," Box 1964A, Entry 420.
5. See article in A.E.F. University News, vol. 1, no. 4, May 15, 1919.
6. This data, submitted by the registrar, R. W. Cooper, is in folder "Statistical Reports," Box 1964, Entry 420.
7. Stars and Stripes, May 9, 1919.
8. Bulletin No. 76, Headquarters, American E.F. University, April 30, 1919, Entry 414.
9. Ibid. This was in response to a memorandum from GHQ, of April 25, to organize homeward bound units, which would go home "upon conclusion of the present term on or about June 15, 1919." The personnel were to be tentatively organized on or before June 1, into special casual companies of not less than 2 officers and approximately 150 men each. The First Replacement Depot at St. Aignan was to send officers to Beaune to see that all records were complete and in order so that the companies could proceed directly from the university to designated base ports. Details are in Bulletin No. 74, Headquarters, American E.F. University, April 28, 1919, Entry 414.
10. Memorandum No. 44, Headquarters, American E.F. University, April 30, 1919, Entry 415.
11. Bulletin No. 76, Headquarters, American E.F. University, April 30, 1919, Entry 414. There is further discussion of these matters in Stars and Stripes, May 9, 1919.
12. Bulletin No. 88, Headquarters, American E.F. University, May 15, 1919, Entry 414; also Bulletin No. 79, Headquarters, American E.F. University, May 6, 1919, Entry 414. The latter stated that, owing to the scarcity of musical instruments, classes scheduled in violin, piano, and violoncello had been canceled.
13. Article in A.E.F. University News, vol. 1, no. 5, May 22, 1919. The Violin Department was under the direction of Walter Karlos Hawkinson, Musician 1st Class, who had prior experience in both orchestra and solo work in the United States.
14. Memorandum No. 44, Headquarters, American E.F. University, April 30, 1919, Entry 415. See also Memorandum No. 48, Headquarters, American E.F. University, May 5, 1919, Entry 415, ordering all organization commanders to take immediate steps to clothe and equip their men as required by embarkation regulations.
15. Ibid.
16. See memorandum, adjutant general, GHQ, AEF, to Commandant, American E.F. University, May 6, 1919, in folder "Recruiting," Box 1923, Entry 408.
17. Bulletin No. 90, Headquarters, American E.F. University, May 16, 1919, Entry 414. There were opportunities for about twenty-five men in the U.S. Army to remain in France and take up studies at French universities on fellowships provided by the Society for American Fellowships in French Universities. See in Memorandum No. 69, Headquarters, American E.F. University, June 2, 1919, Entry 415. It is not known whether anyone from Beaune applied.
18. Bulletin No. 97, Headquarters, American E.F. University, May 19, 1919,

LAST DAYS

Entry 414.

19. Articles in *A.E.F. University News*, vol. 1, no. 5, May 22, 1919. The senior chaplain of the university was Gregory O'Brien, who held the rank of captain.

20. Memorandum No. 54, Headquarters, American E.F. University, May 12, 1919, Entry 415.

21. Memorandum No. 65, Headquarters, American E.F. University, May 26, 1919, Entry 415. The memo also states: "The practice of officers and enlisted men visiting this office, without proper authority, requesting information regarding papers, must be discontinued."

22. Memorandum No. 59, Headquarters, American E.F. University, May 19, 1919, Entry 415.

23. Memorandum No. 61, Headquarters, American E.F. University, May 21, 1919, Entry 415.

24. Memorandum No. 62, Headquarters, American E.F. University, May 22, 1919, Entry 415. See this for list of "authorized and necessary equipment."

25. Bulletin No. 120, Headquarters, American E.F. University, June 4, 1919, Entry 414. The commanding officers at Brest, St. Nazaire, and Marseilles, and of the First Replacement Depot, at St. Aignan, were so informed.

26. Memorandum No. 57, Headquarters, American E.F. University, May 16, 1919, Entry 415.

27. Memorandum No. 68, Headquarters, American E.F. University, May 29, 1919, Entry 415.

28. Article, *A.E.F. University News*, vol. 1, no. 6, May 30, 1919.

29. Ibid.

30. Memoranda Nos. 69 and 70, Headquarters, American E.F. University, June 2 and 3, 1919, Entry 415.

31. General Orders No. 40, Headquarters, American E.F. University, May 16, 1919, Entry 412; Bulletin No. 117, Headquarters, American E.F. University, June 2, 1919, Entry 414.

32. Bulletin No. 122, Headquarters, American E.F. University, June 9, 1919, Entry 414.

33. Memorandum, No. 66, Headquarters, American E.F. University, May 27, 1919, Entry 415. A modified clothing and equipment list was published at that time as well.

34. General Orders No. 44, Headquarters, American E.F. University, June 4, 1919, Entry 412.

35. *A.E.F. University News*, vol. 1, no. 7, June 6, 1919.

36. Ibid.

37. Articles ibid., vol. 1, no. 6, May 30, 1919; Bulletins nos. 104 and 118, Headquarters, American E.F. University, May 22 June 3, 1919, Entry 414.

38. *A.E.F. University News*, vol. 1, no. 7, June 6, 1919.

39. Ibid. The conference was opened with a rural life pageant.

40. Ibid., vol. 1, no. 6, May 30, 1919.

41. There is much discussion of this conference in *Stars and Stripes*, April 25, 1919.

42. *A.E.F. University News*, vol. 1, no. 7, June 6, 1919.

43. Articles, ibid., vol. 1, nos. 4 and 6, May 15 and 30, 1919.

44. See in *Stars and Stripes*, June 6, 1919.

45. Ibid., May 2, 1919.

46. *A.E.F. University News*, vol. 1, no. 7, June 6, 1919.

47. Ibid., vol. 1, no. 5, May 22, 1919.
48. Ibid.
49. Ibid., vol. 1, no. 6, May 30, 1919.
50. Ibid., vol. 1, no. 4, May 15, 1919
51. Ibid., vol. 1, no. 5, May 22, 1919.
52. Lengthy article ibid., vol. 1, no. 6, May 30, 1919.
53. Ibid., vol. 1, no. 7, June 6, 1919. After the war, Reeves established his own engineering firm in Tulsa, Oklahoma. Erskine resumed his distinguished career at Columbia, and continued his writing, publishing numerous volumes of poetry, essays, and literary criticism.
54. General Orders No. 45, Headquarters, American E.F. University, June 6, 1919, Entry 412.
55. Ibid., vol. 1, no. 6, May 30, 1919.
56. Ibid., vol. 1, no. 7, June 6, 1919.
57. There were nearly four hundred of these people then actively engaged throughout the AEF. See article, ibid., vol. 1, no. 6, May 30, 1919.
58. General Orders No. 41, Headquarters, American E.F. University, May 26, 1919, Entry 412. This indicated that the reports from each college would be in the hands of the educational director on June 1; all examination reports to be handed in to the registrar by June 5. After 6:00 P.M., on June 7, all the clerical staff was to be placed at the disposal of the military organization to hasten preparations for departure.
59. *A.E.F. University News*, vol. 1, no. 7, June 6, 1919.

CHAPTER 10:
Conclusions

GENERAL PERSHING once advised the men of the AEF shortly after the armistice, that, "since we did not need to die for America, let us live for her."[1] Nowhere was his injunction more enthusiastically endorsed than by those fortunate members of the AEF who were able to continue their higher education while awaiting shipment home. Erskine himself expressed what he and other leaders had in mind at Beaune: "We intend here to save for some good use if possible the time that would otherwise be spent in irksome waiting for the ship that is to take us home." Therefore, "we intend . . . to seize out of the very handicaps and necessities of the moment some lasting advantage."[2]

The university was partially a reflection of higher education structures and methods then in place in the United States. It grew out of the insight and wisdom of many leaders within and outside of the army: the secretary of war, Newton Baker; Pershing; and highly-placed YMCA officials giving literal meaning to the concept of the army as a "university in khaki." To be sure, it stemmed in part from the necessity of keeping the men of the AEF occupied while awaiting shipment home. But it also sprang from American attitudes and practices regarding education, which, between 1900 and 1920, had experienced a revolution, partially driven by the ideals of Progressivism.

One of the key figures of the school, Dr. John Erskine, exemplified these views. Writing to Professor A. H. Thorndyke of the Department of English at Columbia in May 1918, Erskine asserted that "there never was such a chance to enlarge the moral horizon of a great number of our countrymen." They were not only already studying languages and other subjects, but in various locations were using their leisure to study local history and architecture. Often the camps were near interesting cities or châteaux, "and the men show most astonishing interest in the country." They were becoming quite curious about their own as well, because the French often asked them embarrassing questions about their nation's political and social affairs. Thus, "for purely patriotic reasons the boys would like to have an answer."[3]

The YMCA and the army bent every effort to capitalize upon this favorable educational climate. Indeed, Rees, Reeves, Erskine, and many others, hoped for nothing less than revolutionary reform in higher education, as well as in vocational and technical training, all in a rather unlikely venue, the United States Army.

One innovation at the university at Beaune illustrates this attitude. Making a virtue out of necessity, the notion was advanced that the school should foster open enrollment because the men were more mature than were the usual college students, if not in years, then because of their experiences in a world at war. That encouraged the idea that the citizen should continue to study "in some way or other all of his life." Indeed, as Erskine observed, "one of the shortcomings of education heretofore has been that it has stopped when men and women have begun their life work—at the very moment, that is, when they are likely to find out what they most need to study."[4] The remedy seemed to be an open enrollment policy. Of course, it was manifestly impossible to evaluate properly each student's qualifications to enter the university at Beaune on such short notice and on the scale required. The criteria were therefore altered: "the only test [was] the student's ability to carry the work." If the student could do the work, then let him proceed; if not, he could still obtain educational opportunities. He could enroll in preparatory work or study a technical field in the model divisional school at Beaune. Or art students at Bellevue could elect to take more general, less demanding, art classes at Beaune. Students interested in agriculture could attempt to pursue knowledge in the field in the College of Agriculture at Beaune, or choose the more elementary, general course at Allerey. Indeed, Erskine and others saw the concept of open enrollment as a beacon to the schools at home, predicting "that this policy will come into more general adoption in the United States."

The AEF University was breaking new ground in another way. At Beaune, officers, enlisted men and civilians shared the teaching responsibilities, which had interesting ramifications. As Erskine noted, "one of the most fruitful aspects of our work here is that the teaching is done by fellow citizens, that is by officers and men who, a short time ago, were military comrades in arms." This was a welcome innovation, he declared, and "we must learn at home not to leave the sole privilege of teaching to the professional educator." What the teaching staff at Beaune was doing was the "learning of a kind of charity which will share intellectual

CONCLUSIONS

wealth as men now share material goods."[5] Perhaps better put, "we are reduced to a society of fellow-citizens, each trying to help the other to a little knowledge."[6] Related to this was another new procedure, "itself a novel undertaking," whereby enlisted men were detailed as instructors, in situations where officers up to the grade of major would be included as students. There had been many who "shook their heads gravely," saying that "it would never do, at all." But no clashes were reported, and all went well, to the credit of all concerned.[7]

The AEF saw itself as the cutting edge in other respects. Regarding the athletic program, both on campus and beyond the university, the ideal was "athletics for everyone." Though the school and other AEF units fielded teams of almost professional quality, the ideal was intramural competition, which should also be emulated at home.

The AEF University set up thirteen traditional academic colleges. Even here, advances were planned or instituted. For example, supporters of business colleges found at Beaune innovations in business instruction, focusing on specialty clubs and encounters with foreign, i.e., French, industrial, commercial, and business methods, gaining some respectability for the discipline.

The College of Education made gains in the model post and division schools on the campus, and in other respects, regarded the AEF as a vast laboratory for the development of adult and continuing education. The college also provided for a large number of illiterates, teaching "English as a foreign language." William S. Coy, the assistant supervisor of instructors of Negro illiterates, had praise as well as criticism for this instruction. He felt that the work had "done an immense amount of good," and that many of the blacks had pledged to continue their education when they returned home. Yet, the work had been handicapped because it had never before been given serious consideration; there was no experience to build upon. Consequently, the teaching of reading, arithmetic, and kindred subjects to illiterate adults was "a field for research entirely new." Certainly, "ordinary primary methods with children in practice for some years now do not sufficiently answer," he declared. Furthermore, there were many lapses and weaknesses, such as a looseness of organization, the frequent changing of instructors, and the tendency of many of the young and untrained teachers to talk too much rather than allow the student time for his own expression. This was necessary, Coy continued, observing that as William James had insisted, for good instruction "there can be no impression, without expression."

Nevertheless, Coy was well pleased that in the post and divisional schools much more than a mere beginning had been made toward a new educational era.[8]

In other areas, the AEF University established colleges that met specific needs within the army. The Cadet College was only for the instruction of candidates for West Point. In the event, it provided the academy with nineteen plebes. The College of Correspondence became the university's extension school and built upon the correspondence school traditions at home. The Department of Citizenship was considered by some commentators as a major advance that should be widely copied by schools in the United States, where civic matters were, more often than not, widely ignored.

The AEF University also demonstrated, to the satisfaction of most, even Pershing, that military discipline need not be detrimental to the pursuit of higher education, even if it did fall somewhat short of fostering the traditional open pursuit of knowledge. the *Stars and Stripes* once observed that the university at Beaune gave "direct and tangible proof" that the academic and military life could be successfully combined.[9] Pershing, in his final report, agreed, noting that in general the AEF's educational programs had indicated that discipline had not suffered appreciably.[10]

Nor did Beaune lack that accoutrement of traditional American higher education institutions, the elusive but often pervasive thing known as "college spirit." Its ability to inculcate this is one measure of its success. It was understood that, in the usual sense, college spirit depended on "spreading oaks and moss-covered walls," and an atmosphere of culture, mellowed by time. This was buttressed by athletic records and other traditions. Of course, the AEF University could not count on these things, but it did have ready-made, if recent, traditions and common experiences that its students possessed, such as memories of the trenches and the fighting line, of dugouts and shell-torn shelters, of wearing marches and sleepless vigilance, and of sharing the tasks of war and tasting the victory that resulted. Coupled with this profound sense of comradeship, "there is a keenness for learning so characteristic of this body of students that it has occasioned the comment of prominent visitors to the University," as one writer observed. Thus, the AEF University possessed—the precursor of that body of G.I.'s of the next generation that flocked to colleges and universities following World War II—"a great body of men, bound together by months of hardships and joys, shared alike; men eager to return home and take up anew the duties and privileges

CONCLUSIONS

of citizenship, but even more eager to add to their knowledge and better to fit themselves for citizenship." In such an atmosphere there should surely be a bona fide "college spirit." And so there was. Despite the horrors of the recent conflict, much of America's earlier optimism remained in the hearts of the soldier-students participating in the AEF's remarkable ventures in higher education, and they did not fail to recognize and accept the traditional notion of the college, even the University at Beaune, as their alma mater.[11]

At some point, many of the students at Beaune evaluated their educational experiences. In their unsigned statements a picture of the accomplishments of various educational programs can be obtained.[12] One student, enrolled in a class of forty-five in commercial arithmetic, apparently in the post or division school at Beaune, noted that the physical conditions were deplorable: there was no floor on the classroom except for the wood shavings scattered around. The room's partition, the windows, and the upper part of the door were of cheesecloth. There was only one small blackboard about three feet square and a piece of a towel for an eraser. There was no stove in the room and the weather was cold. The benches and tables were made of rough boards. Yet he concluded: "My experiences with this class were as pleasant as any I have ever had." The class was lively and the instructor competent, and many of the students stated that they "had never had a course anywhere that they had enjoyed more." One thought that education in the army was "a very good thing in view of maintaining a citizen army in the United States." The combination of schoolwork and military training enabled the soldier to serve his country and to benefit himself as well. Another praised Reeves for creating a wonderful organization, and did not think that a more perfect educational program could have been set up in the short time available. He did object, though, to the compulsory attendance at the citizenship lectures. A sergeant remarked that a system of army schools would attract more men into the service, and perhaps encourage those who were in to remain with the colors for longer periods of enlistment. Another soldier-student was certain that "my months of waiting could not have been more profitably spent anywhere in the A.E.F.," than at the university. Indeed, he asserted, "I have increased my knowledge far beyond my expectations." Another praised the institution for helping prepare the men to return to civilian life, the principles of which, "in a few months of army service . . .are forgotten." He concluded that all recruits should attend schools in the army, combining educational pursuits with their military training.

These sentiments were echoed by Erskine, who wanted to build on what the university had achieved. The Beaune experience should be only a beginning, he once affirmed. "It has interested me exceedingly to observe, as I have advanced the idea that our Army should be transformed into an instrument for education, that the idea has met with intelligent response from the experienced officers of the Regular Army." The university had come to mean so much, even in its short life, to the degree that "our most chivalrous soldiers, our best trained scientists, our citizens with the largest imagination may [now] agree on a program of universal training which will quickly drive out illiteracy and all disease that is avoidable, and will fit men and women to take a worthy part in the peace time business of our country." If the United States could be so organized, Erskine was certain that "we should be prepared against attack in time of war, and we should also be prepared, as far as is humanly possible, against those prejudices and ignorances which tend to drive nations into war." What also seemed of major importance was "that this work is carried out within the Army," a fact that "may prove of immense significance." In fact, he went on, "We are using an army for the first time to advance purely peaceful ends. The military organization controls the discipline and furnishes the material resources of the school, but the intellectual life is left absolutely free to adjust itself to modern social tendencies." This experiment might prove the feasibility of introducing into the nation's life "a system of universal training not yet attempted—a system which should be compulsory for all men whether or not they are physically fit to be soldiers." The university at Beaune could serve as a model for the training centers such a system would require. There, a student would be under military discipline, enjoying "all advantages of the well administered military camp," but he could also "pursue whatever subjects his tastes and his situation in life might cause him to desire, from brick-laying to university study." To be sure, Erskine was well aware that the army had as its first priority the necessity to prepare "to fight in a traditional sense, and the need for its discipline would remain," but there was much more to be expected of the armed forces.[13] As Erskine put it elsewhere, "we wish [the university] to stand for the plea of national training." Therefore, "each one of us must see that our country takes its part by organizing what might be called a national army against ignorance."[14] To attain these ends, he suggested that the War Department be combined with an Educational Department, forming a new Department of National Defense. In this way, the nation could be organized "as effectively against disease and ignorance as it has been against

CONCLUSIONS

less subtle enemies." Undoubtedly, "the way to save us from the dagger of militarism, . . . is to develop the chivalry of the soldier's career in other directions than mere fighting."[15]

Members of the Y's Army Educational Corps Commission at Beaune concurred, suggesting that Erskine's proposed schools, sustained by a year-long, peacetime draft, might be called National Civic Institutes. These could, in addition, perhaps help in achieving the ends of a genuine national unity, which "is still before us" to be achieved, the war's crisis having "disclosed how far we are from this goal, and brought home to us the supreme importance of attaining it." At the end of each year's educational and training cycle in the institutes, the nation would add more than a million men to its resources, trained not strictly in military affairs, but equally in civic knowledge and ideas, and in many of the arts and industries that would contribute to the prosperity of the country at peace, as well as produce many of the skills that had been found indispensable during the war.[16]

Thus it was difficult for many Americans, notably those in the AEF's educational programs and leaders in the Y, to relinquish the spirit of America's wartime Great Crusade with its progressive underpinnings. Encouraged by what had been accomplished, which was, after all, considerable, they were perhaps deluded about its significance and possibilities, and naive about its lasting effects. Finally, it should be noted that these high resolves were not in harmony with the public's and the AEF's more mundane momentary aspirations: to get the men home; to forget all about their overseas ventures; and to get on with "normalcy." In this respect, the army was ahead of its time.

It is interesting to compare Erskine's educational ideas with those of a later twentieth-century commentator, Harry G. Summers, Jr., who, in his book *On Strategy II. A Critical Analysis of The Gulf War* expresses similar ideas regarding the army's future: "But while training will remain important in the post-Cold War world, education must now be given priority in the military's staff colleges and war colleges. One can train for the known, but only education can prepare the military for the unknown and unknowable futures that lie ahead." Therefore, not only must equipment research and development be continued beyond the Cold War era, but in his view, "even more important will be the development of the flexibility of mind" among servicemen and women, "necessary to cope with the unexpected."[17]

But if many of the visions of Erskine and others did not come to pass, what, then, were the legacies of the AEF's educational ventures?

Regarding the school at Beaune, certain short-term accomplishments seemed noteworthy. To a writer in the *A.E.F. University News*, the testimony and proof of its solid accomplishments were "so evident as to make comment superfluous." The results would be lasting; students would carry the seeds of this success across the sea "and into the hundreds of activities that its students have been better fitted to pursue by their work in Beaune." However, this success had not been accidental. It resulted from the vigor, initiative, decision, and rugged strength of character of the leadership, which had overcome myriad obstacles in bringing the school into existence. Some of the men involved were initially appalled by the task's "very bigness and the difficulties attending its execution," but even the faint-hearted soon fell to work and forgot their doubts. Indeed, almost immediately "they saw buildings, occupied by classes, on spots that a day or two before had been open ground: they saw men thronging a library that has had to be enlarged four times to accommodate its public—in short, they saw accomplishment and success."

They found a ready response from the doughboys. In the main, these men, expecting so soon to return to civil life, had a deep hunger "for knowledge that would best fit them for chosen tasks." Therefore, "developed in incredibly brief time, shaped by a few earnest men into a sensate, pulsing, powerful institution, the University is an exemplar of American initiative and ability." It remained "the crowning educational achievement of an Army that has devoted unprecedented attention to education."[18]

Erskine, in a report to Frederick P. Keppel, at the War Department in Washington, elaborated. He noted that, in the first place, the change it had made in the spirit of the men was enormous; they were "extremely happy and in good humor." And though there were other benefits, this fact alone would justify all of the time, expense, and effort expended.[19]

Therefore, in the short term, the aspirations of many in America were realized by the AEF's educational initiatives. Something good had come from the war, which had helped "pay for itself by creating a better America," if not a better world.[20] In this sense, as one writer has observed, "the American war effort was the climax of the Progressive movement."[21]

If the AEF's educational ventures did not come out as many hoped for, especially regarding a continuation of some sort of traditional higher education programs under army auspices, there was another legacy.

There is little doubt that General Pershing continued to hold to the

CONCLUSIONS

view that the army was a major teacher. The historian and journalist Mark S. Watson, who had been one of the officers in charge of the *Stars and Stripes* in Paris during World War I, and was later a Pulitzer prize-winning writer for the *Baltimore Sun*, once observed that when Pershing was chief of staff from 1921 to 1924, "he did more than any other man since [Elihu] Root to build up the school system for the Army." He had in mind the branch, technical, and service schools, which, with the Command and General Staff School and the Army War College, all provided the hard core of military talent that contributed so much to America's military successes during World War II. The school system, Watson was certain, was "Pershing's most enduring gift to the country and his most enduring monument."[22]

General Eisenhower concurred. In a cable to Pershing following Germany's surrender on May 8, 1945, he emphasized Pershing's contributions to the victory. Recognizing that his work in reorganizing and expanding the army school system was a major achievement, Eisenhower concluded: "The stamp of Benning, Sill, Riley, and Leavenworth is on every American battle in Europe and Africa. The sons of the men you led in the battle in 1918 have much for which to thank you." He might have added that these accomplishments clearly owed much to the school innovations and initiatives undertaken on behalf of the AEF, which grew out of Pershing's convictions about education, coupled with the immediate need to keep the men busy, a nice dovetailing of needs, hopes and aspirations, fateful indeed.[23]

Pershing himself was aware of the significance of army education, as is revealed in his brief remarks on the occasion of the great victory parade held in Washington, D.C., on September 17, 1919, as the nation's capital welcomed Pershing and many of his soldiers home from the wars. At that time, Pershing stated that his men had "returned in the full vigor of manhood, strong and clean. In the community of effort, men from all walks of life have learned to know and to appreciate each other. . . . They will bring into the life of our country a deeper love for our institutions and a more intelligent devotion to the duties of citizenship." He obviously had in mind many of the programs that he had developed—indeed insisted upon—in the AEF, such as the rigorous anti-venereal disease campaign, and the educational initiatives that he had warmly embraced.[24]

Consequently, even though Pershing and others could not convince the American people that some sort of continuing universal military training was necessary, and the visionaries at Beaune could arouse no enthusiasm for the civil institutes, Pershing did carry forward the educational innovations to the extent possible given the climate of opinion (and chances of financial support) in the nation. There would be no equivalent of the G.I. Bill following World War I. But in a limited way, within the army, educational initiatives were undertaken that would at least permit it to develop further in its primary military roles; what the army and American society could and would do for the citizen-soldier would have to wait for further developments, leading at last to a climate of opinion that would accept, perhaps even demand, the initiatives of the G.I. Bill, and, later, enhanced training and educational programs within the armed forces.

Although there was no G.I. Bill following World War I, the YMCA expended over six million dollars to further its educational aims in assisting ex-servicemen. Of these funds, five million dollars were allocated to scholarship awards, for any level of schooling from elementary through university, to assist discharged veterans desiring education or training. In addition, at the elementary level, the Y set up numerous schools of its own, especially to assist illiterates, and a program of correspondence courses was instituted. In all, almost ninety thousand scholarships were awarded, over twelve thousand at the college and university level. Over twenty-four thousand students enrolled in correspondence courses. The Y also continued to foster instruction in citizenship. Lectures and institutes for these ends were often conducted in local posts of the American Legion, which was being organized during this time.[25]

Pershing, in his final report to the War Department, summarized the quantitative aspects of the educational ventures of the AEF. His enrollment figures are rather low regarding the higher educational ventures. He recorded that the university at Beaune had enrolled 8,528 students at its peak, with 367 more being accommodated at the Art Center. His final figures pertaining to the AEF's educational endeavors included 4,144 mechanical trade school students, 6,300 doughboys attending various French universities, and 1,956 in similar institutions in Britain.[26] Together with 690,000 enrollees in institute short courses, 750,000 attending extension lectures, 181,475 in the post schools, and 27,250 in the divisional schools, a grand total of 1,670,020 AEF officers and men participated in some way in the AEF's educational programs. All of this "demonstrated satisfactorily that a combined military and educational program can be carried out in the Army with little detriment to pure military training and with decided advantage to the

CONCLUSIONS

individual," he concluded.[27]

Altogether, the educational initiatives of the AEF were a substantial experiment, which has unaccountably received little attention. They suggest what enlightened military leadership is capable of producing; swords can be turned into plowshares. The army had not only emerged as a chief instrument in the crusade "to make the world safe for democracy," it helped to shape that democracy. The army demonstrated that it had more to do than provide the machinery for inducting, training, bringing to battle, and then discharging its personnel. It could be transformed into an instrument helping to create a new world. If far removed from such modern initiatives as the American Military University and its "distance learning" opportunities and innovations, certainly the university at Beaune was an advanced educational endeavor for its time, a not inconsequential forerunner of numerous programs that are now established parts of the military.

Of the AEF's two million men, about twenty thousand or about one percent—less than the strength of one of its wartime divisions—received unparalleled, superb opportunities to study abroad at the university level. This marks perhaps the finest hour of the YMCA, and of such military leaders as Erskine, Reeves, and Rees. Dr. Kenyon Butterfield once observed that "the time will come when the achievement of the American Army in building a University . . . in the space of a month will be written into the history of education as one of its most unique paragraphs." What was significant was not the GHQ's plan "for the sake of keeping men out of mischief," but the high command's glimpse of much greater possibilities. America had emerged as a great power "as a result of the demonstration of its effectiveness in this war." But this created responsibilities and obligations. The nation now had to find and educate the leaders to assume the new economic, political, and moral leadership roles it had acquired. Those leaders must be trained, "both into a clear insight of the problems at stake and effective capacity for fulfilling the tasks that are thus imposed." The university at Beaune, "on the soil of France, which America has helped to rescue from destruction," had begun the task of helping students and teachers alike to cooperate "in an endeavor to arrive at new power to meet the issues of a new day." This, in the last analysis, was, in the view of one of its more thoughtful mentors, "the real meaning of the American E.F. University," serving as well to describe the importance of all of the AEF's educational initiatives.[28]

Following the armistice, at a time when it seemed that they would never return home, under the leadership of some remarkable educators, both within and outside of the army, American soldier-scholars embarked upon some rather unlikely academic ventures. Throughout Great Britain and France, but especially on the plains of Burgundy at Beaune, in the Côte d'Or, they descended in substantial numbers. In often less than idyllic conditions, they exemplified the ideals of Chaucer's wandering clerk-scholar of Oxford,

>And gladly wolde he lerne,
>and gladly teche.

CONCLUSIONS

NOTES

1. *Coming Back*, a newspaper published by the National War Work Council of the YMCA for the benefit of returning doughboys, nos. 7 and 9, February 14 and 28, 1919.

2. Published address by John Erskine, *Society As A University* (Dijon: R. De Thorey, 1919), p. 3.

3. See letter, Erskine to Prof. A. H. Thorndyke, Department of English, Columbia University, May 21, 1918, in undesignated folder, Box 1965, Entry 420.

4. Also as Erskine in another place expressed it: "May we learn here and carry back home with us the important truth that no man should ever consider himself beyond the school age." See Erskine, *Society As A University*, p. 3.

5. See memorandum, "Notes Of The University," by John Erskine, April 27, 1919, in folder "Reports," Box 1965, Entry 420.

6. Erskine, *Society As A University*, p. 3.

7. Two articles in A.E.F. *University News*, vol. 1, no. 6, May 30, 1919.

8. See collection of comments in folder "Reports—Work in Classes by Instructors," Box 1930, Entry 419.

9. *Stars and Stripes*, May 16, 1919.

10. Pershing's remarks in his final report in *United States Army in the World War*, vol. 12, p. 69.

11. *Bulletin of The A.E.F. University News*, vol. 1, no. 3, May 9, 1919.

12. See collection of student comments in folder "Reports—Work in Classes by Instructors," Box 1930, Entry 419.

13. Memorandum "Notes Of The University," by Dr. John Erskine, April 27, 1919, in folder "Reports," Box 1965, Entry 420.

14. Erskine, *Society As A University*, p. 11.

15. Letter, Erskine to Frederick P. Keppel, third assistant secretary of war, March 19, 1919, in folder "Publicity and Public Press," Box 1901, Entry 408.

16. Lengthy article in *Stars and Stripes*, May 23, 1919.

17. Harry G. Summers, Jr., *On Strategy II. A Critical Analysis of The Gulf War* (New York: Dell, 1991), p. 266.

18. A.E.F. *University News*, vol. 1, no. 4, May 15, 1919.

19. Letter, Erskine to Frederick P. Keppel, third assistant secretary of war, March 19, 1919, in folder "Publicity and Public Press," Box 1901, Entry 408.

20. Baldwin, "The American Enlisted Man," p. 235.

21. Ibid., p. 236.

22. See article, "Pershing—Nation's Greatest Soldier," *Baltimore Sun*, August 28, 1960, as quoted in Donald Smythe, *Pershing: General of the Armies* (Bloomington: Indiana University Press, 1986), pp. 280, n. 28; 350. Smythe also notes that Pershing was the first army chief of staff to get legislation passed allowing the sending of officers to private schools for special studies, as, for example, enabling him to send his aide, J. Thomas Schneider, to Harvard for a law degree. Becoming secretary of war in 1899, Root became well known for many reforms in the army, many involving education and training.

23. As quoted ibid., p. 306. The schools that Eisenhower mentioned were respectively: infantry, artillery, cavalry, and command and general staff. However, there is no doubt that the U.S. Army had already instigated the substantial development of the professional schools before Pershing's enhancement programs, even before World War

SOLDIER-SCHOLARS

1. Edward M. Coffman has convincingly argued that Leavenworth's School of the Line and its staff college were substantially reformed after the Spanish-American War. In addition, the Army War College was founded in 1903, further enhancing the army's educational establishment. See his "The AEF Leaders Education for War," in R.J.Q. Adams, ed., *The Great War, 1914-18: Essays on the Military, Political and Social History of the First World War* (College Station: Texas A&M University Press, 1990), pp. 139-59. Thus Pershing had a substantial foundation upon which to build.

24. See account of the Washington, D.C., victory parade and Pershing's address on that occasion in Smythe, *Pershing*, p. 261.

25. Taft, *Service With Fighting Men*, 1: 360-63.

26. *Stars and Stripes*, February 21, 1919, published some preliminary figures on enrollment of American soldiers in French universities.

27. See final report of General John J. Pershing, to secretary of war, from GHQ, AEF, Paris, September 1, 1919, in *United States Army in the World War, 1917-1919*, vol. 12, p. 69. Pershing's entire report is on pp. 15-71.

28. "A Word From Commissioner Butterfield," March 10, 1919, attached to address by Erskine, *Society As A University*, pp. 11-12.

Index

Abbott, Second Lieutenant Edmund Quincy, 51

A.E.F University News, student newspaper of the American Expeditionary Forces University, Beaune, France, 61, 72-73, 76, 186

Aix-les-Bains, France, 3

Ajaccio, Corsica, 155

Alaux, M., 111

Albright, Roy, 40

Alexander, First Lieutenant Horace K., 54

Alpine-American, student newspaper, University of Grenoble, 153

American Dijonnais, student newspaper, University of Dijon, 153

American Expeditionary Forces [AEF], xii, xiii, xv, 1;
  agricultural study in, 83-84;
  discipline in, 1; drill in, 1;
  education in, 6, 7, 8, 9, 10;
  leaves in, 2, 3;
  soldier shows in, 5;
  sports in, 3-4, 134-35

American Expeditionary Forces Art School, Bellevue, France, 97-119;
  courses at, 108-15;
  establishment of, 105-09;
  faculty of, 109-12;
  results of, 116-19

American Expeditionary Forces College Debating League, 152

American Expeditionary Forces Entertainment Bulletin No. 1 (GHQ), 4

American Expeditionary Forces School of Architecture (Le Mans, France), 104

American Expeditionary Forces University, Beaune, France, xii, 10;
  army regulations and, 18-19, 21;
  athletics at, 74-75;
  attitudes of students toward, 182-83;
  band of, 40;
  blacks in, 32, 33, 181;
  college spirit at, xii, 182;
  discipline at, 35-36, 37-38, 39-40, 72-73, 167-68;
  founding of, 14, 15, 17, 19, 21;
  fraternities in, 18;
  military organization of, 22-23;
  opening of, 34-35;
  publicity at, 76;
  schedules at, 21, 24-25, 34, 40, 41;
  University Council of, 19, 21, 60;
  women in, 35, 65

American Institute of Architects, 104

American Jewish Chronicle, 60

American Legion, xiv, 188

American Library Association, 19, 25, 27, 87, 110

American Medical Post-Graduate Society, 133

American Red Cross, 27, 35, 66, 71, 97, 102, 106, 109, 114, 125

American Soldier-Student, student newspaper, British detachment, 137

American University Union (Paris), 7, 130, 133, 138, 141

Amherst College, 31

Anglo-American Committee of the Fellowship of Medicine, 133

Anglo-American Relations, 130-38

Anti-tuberculosis Society of Beaune, 126

Archer School of Modern Piano Playing (Pittsburgh), 64

Argonne, Battle of, 73

Army Educational Commission (YMCA), 7, 8, 19, 21, 24, 32, 57, 77, 82, 83, 97, 98, 99, 130, 138, 170, 171

Articles of War, 72

Associated Press, 60

Association of Public Utility, 125, 126

Astor, Mrs. Waldorf, 133

As You Were, student newspaper, University of Rennes, 152

Atlanta Conservatory of Music, 65

Atterbury, Grosvenor, 30

Aubert, Louis, 174

Auvergne, region of France, 3

Babcock, Major Franklin, 21

Baillot, M., 142

Baker, Herbert J., 84

Baker, Newton Diehl, U.S. Secretary of War, 6, 74, 179

Balfour, Arthur James, British Foreign Secretary, 133

Baltimore Sun, 187

Bandmasters School (Chaumont, France), 118

Banks, Major L. Frazer, 32, 33

Barnum, Major F. E., 23

Bastel, Private First Class Frederick E., 64

Baylor University, 60

Bayne, First Lieutenant Thomas S., 86

Beaune Committee of French Homes, 37, 170

Beaux Arts Institute of Design (New York City), 99, 119

Beilby, Private First Class Smith G., 86

Belleau Wood, Battle of, 22

193

Benson, First Lieutenant Charles C., 37
Besançon, France, 51
Besançonian, student newspaper, University of Besançon, 153
Biarritz, France, 3
Birkbeck College, 130, 132
Birmingham (England) *Mail,* 132
Black, First Lieutenant Robert S., 106
Black, Hugh, 42
Black, Second Lieutenant John Earl, 56
Blarney Castle (Cork, Ireland), 133
Bonner, Captain L. A., 23, 38
Bordeaux, France, 83
Borglum, Gutzon, 109, 116
Borglum, Solon Hannibal, 109, 114
Boston Transcript, 60
Bouze, France, 41
Bowdoin College, 131
Branch, Private Robert L., 28
Brest, France, 83
British-American Club (Oxford University), 133
Brooke, Major Richard, 24, 174
Brown, Captain Thomas Stephen, 54
Brown, Elwood S., 4
Brown, Ensign Archibald M., 99, 109, 111
Brown University, 59
Bulletin of the A.E.F. University News, 61, 76
Bush, Major Andrew J., 17, 23
Butler, Parker, 72
Butterfield, Kenyon Leech, 7, 57, 82, 83, 84, 87, 174, 189

Cadet College (AEF University, Beaune), 25-26, 50, 182
Cambridge University, 10, 130, 134
Canadians, 10
Canal Zone, 22
Cannon, Nan, 160
Carcassonne, France, 157
Carlu, M., 111
Carrière, Eugenè, 112
Carrière, Jean-René, 112
Carroll, Captain M.C., 73
Carroll, First Lieutenant James B., 114
Catts, Lieutenant Colonel Gordon R., 142, 157-58, 160
Cauldwell, Captain Leslie, 109, 110, 112
Cellarius, Lieutenant Charles, 99, 102
Chaillaux, Lieutenant Homer L., 106
Chambellan, Sergeant René P., 102
Chamonix, France, 3
Chaucer, Geoffrey, 190
Chaumont, France (General Headquarters, AEF), 1, 4, 7, 9, 10, 14, 15, 28, 36, 42, 78, 82, 118, 141, 142, 160
Chicago Tribune, 76
Chicago Tribune (Paris edition), 149
Church, Private M.M., 28
Citizenship Course (AEF University, Beaune), 21, 58; operations of, 29-31, 34; organization of, 28-29
Citizenship Department of the United States Army Educational Corps, 9
Claveille, M., 173-74
Clermont-Ferrand, France, 3
Coast-to-Coast Company (theatrical group), 172
Coblenz, Germany, 1, 25, 99, 104
Cold War, xv, 185
College of:
    Agriculture (AEF University, Beaune), 17, 57, 84, 87, 90, 92, 172, 180
    Business (AEF University, Beaune), 5-51, 53;
    clubs in, 51-52
    Dentistry (AEF University, Beaune), 54, 55
    Education (AEF University, Beaune), 32, 33, 61-62, 181
    Engineering (AEF University, Beaune), 56-57
    Fine and Applied Arts (AEF University, Beaune), 17, 38, 55- 56, 99, 173
    Journalism (AEF University, Beaune), 39-40, 59-61, 76
    Law (AEF University, Beaune), 42, 58
    Legal Education (London), 130
    Letters (AEF University, Beaune), 60, 62-63
    Music (AEF University, Beaune), 63-65, 126, 167
    Pharmacy (AEF University, Beaune), 54, 55
    Science (AEF University, Beaune), 28, 54 55, 58-59, 63
    Veterinary Medicine (AEF University Beaune), 54, 55
Colliers' Weekly, 149
Columbia University, 7, 19, 60, 64, 98, 131, 141, 179
Command and General Staff School, 187
Committee of Devastated Areas, 126
Committee on Excursions, 75
Conservatory of Music (Fontainebleau Palace, France), 118
Coolidge, Calvin, xiv
Cooper, Richard Watson, 43
Cormon, M., 113
Cornell University, 85, 86, 87
Correspondence College (AEF University, Beaune), 26-28, 50, 182

INDEX

Côte d'Or, area of France, xii, 10, 190
Courtois, Du, 173
Coxhead, Ernest, 104
Coxhead School (Le Mans, France), 104, 105
Coy, William S., 181, 182
Crimanelli, Georges, 73
Cuba, 14, 23

Daily Mail (London), 76
Daily Press (Asbury Park, New Jersey), 59
Dallam, Colonel S. Field, 22-23, 92
Damrosch, Walter Johannes, 118
Danforth, Second Lieutenant George N., 85
Daniels, Josephus, U.S. Secretary of the Navy, 6, 132
Dappert, First Lieutenant Anselmn Fulton, 64
Dartmouth College, 22
Davis, A.C., 33
Dawson, John Charles, 142
Decizè, France, 9
Demotte, M., 110
Delamare, René M., 115
Delaware Agricultural College, 57
Dennis, Second Lieutenant Clifford E., 86
Department of the Fine and Applied Arts (YMCA), 98, 102
Department of Citizenship, 182
Department of National Defense, 184
Department of University Extension, State Board of Education, Commonwealth of Massachusetts, 27
Deux Mots, student newspaper, University of Clermont-Ferrand, 153
Dewey, John, 6
Dickerson, Luther L., 25
Dickson, Private James C., 28
Dijon, France, xii, 24, 55, 60, 76
Disabled American Veterans, xiv
Division Schools (AEF), 31-32, 33, 35
Drew, First Lieutenant Herbert E., 86
Drum, Hugh, 43
Duncan, Isadora, 97, 106

Ecole des Beaux-Arts, 98, 99, 102, 109, 111, 113, 119
Educational Sub-Section, G-5, GHQ, Chaumont, 14
Eighth Cavalry, 23
Eight Hundred and Seventh Pioneer Infantry, 22, 86
Eighty-Eighth Division, 5

Eighty-First Division, 85
Eighty-Ninth Division, 84
Eisenhower, General Dwight David, 187
El Caney, Cuba, 14
Elder, Thomas E., 85
Eleventh Regiment Bulletin, student newspaper, Farm School, Allerey, 91
Embury, Captain Aymar, II, 99, 102
Emporia (Kansas) Gazette, 41
English-Speaking Union, 133
Erskine, John, 7, 19, 30, 42, 97, 127, 138, 140, 171, 174, 179, 180, 184, 185, 189
Exton, Colonel Charles W., 159

Farm School, Allerey, xii, 22, 57, 74, 75, 82-93, 168-69, 171-72, 180;
    aims of, 87-88;
    athletics at, 91;
    clubs at, 89;
    courses at, 85-88, 90;
    drills at, 90;
    end of, 92-93;
    entertainment at, 90-91;
    excursions at, 89-90;
    faculty at, 86-87;
    founding of, 82-86;
    life at, 88;
    student newspapers at, 91-92
Fenski, Private H., 28
Fifteenth Cavalry, 22, 24
Fifteenth Engineers, 56
Fifth Army Corps, 23
Fifth Infantry Regiment, 60
First Army Corps, 23
First Cavalry, 23
First Division, 107
First Vermont Infantry, 15
Five Hundred and Sixteenth Engineers, 106
Florida State University, 86
Fogg, Miller Moore, 39, 59, 76
Fontainebleau Palace, France, 118
Fontainebleau School of the Fine Arts, 118, 119
Ford, Colonel Joseph Herbert, 23, 31, 54
Forestier, J.C.N., 110, 111
Fort Benning, Georgia, 187
Fort Leavenworth, Kansas, 187
Fort Riley, Kansas, 23, 187
Fort Sill, Oklahoma, 187
Fort Worth Record, 60
Foster, Herbert D., 63
Foster, Lieutenant William D., 99, 105
Foster, Private L.P., 86
Foyers du Soldat, 19, 109, 125
Franco-American Club, 63

Franco-American relations, 41, 72, 90, 123-27, 141-42, 146, 147-51, 157-58, 159-60, 173-75
Frank, First Lieutenant Leroy R., 86
Fraser, Private First Class Wilson Poitevent, 64
French-American Club (Beaune), 173
French Central Laboratory (Dijon), 54
French Homes Committee, 170
French, Lieutenant John, M., 32
French Red Cross, 106
Fritsch-Estrangin, M.H., 110

Gasser, First Lieutenant Fred, 168
Gassman, Private Charles P., 28
Geary, First Lieutenant Robert E., 51
General Orders No. 6 (GHQ), 19, 21
General Orders No. 9 (GHQ), 8, 31, 77
General Orders No. 14 (GHQ), 3, 21
General Orders No. 30 (GHQ), 9, 10, 14, 77
General Orders No. 68 (GHQ), 78
General Orders No. 192 (GHQ), 8
General Orders No. 207 (GHQ), 1
General Orders No. 241 (GHQ), 2, 3, 4
German prisoners-of-war, 24
Gheen, Captain Russel T., 85
G.I. Bill, xv, 188
Gièvres, France, 9, 106
Giraldon, Adolphe, 110
Gofnell, Captain E.P., 24
Goodnight, Farrier Clarence L., 86
Goodrich, F.L.D., 25
Gray, Major George H., 97, 99, 105, 107, 108, 109, 110, 111, 116
Gregg System of Shorthand, 28
Grinnell College, 25
Gusman, Captain Charles S., 106
Gusman, Pierre, 113
Gwinn, Joseph M., 61

Haffner, J. J., 10
Haines, Private Clarence S., 64
Hair, Captain Waldo P., 22
Hamilton, Captain Rubey J., 85, 86
Hartford Art Society, 109
Harvard University, 59, 86, 141
Hayward, Harry, 57
Hearn, Private First Class Edward French, 64
Hebrard, Jean, 56, 102
Hellman, George Sidney, 55, 97-98, 99, 102, 105, 108, 117, 118, 173
Hemingway, Ernest, 117
Henderson, Captain Isham, 26
Hill, J. Forest, 26, 28

Hisketh, Captain, 23
Homer, Eleazer Bartlett, 56, 99
Hoover, Herbert, 31, 170
Hôtel de la Poste, Beaune, 54, 125, 174
Howard, Captain Clarence E., 109
Howard College, 142
Howard, John Galen, 56, 102, 110
Howard, Major William N., 22, 62, 85
Howell, Willey, 43
Huff, Captain E., 40
Hurley, Major Patrick J., 22

Illinois College of Agriculture, 85
Inns of Court, xii, 131
Institute of Musical Art (New York City), 64
Inter-Allied Conference on World Agriculture, 92, 170
Inter-Allied Games, 113, 114, 170
International Correspondence Schools, 26
Iowa State University, 60, 86

Jackson, Lieutenant Colonel John Price, 30
James, William, 181
Jewish Welfare Board, 71
Johnson, Colonel Wait C., 4
Journal de Beaune, 72
Judd, Major A.C., 84

Kansas Agricultural College, 57
Kansas City Journal, 60
Kansas City Star, 60
Kaufmann, Reginald Wright, 41, 61
Kelly, Colonel John R., 4
Keppel, Frederick P., 98, 186
Kerlin, Robert T., 86
"Khaki Colleges" (Canadian), 10
Kingsbury, John A., 172
Knights of Columbus, 71, 74, 173

Labrely, Galotin, 73
Lachman, Harry B., 113
Lafferre, M., 173
Laloux, Victor, 111, 119
Landry, Second Lieutenant John Kunkelman Miller, 64
Lane, Alfred C., 41
League of Nations, 40, 153, 171
Leland, First Lieutenant Harold L., 106
Leland Stanford Junior University, 28
Le Mans, France, 83, 99, 104, 105, 174
Le petit méridional (Montpellier, France), 152
Les Beaux Jours, student newspaper, University of Poitiers, 142, 152, 153, 155
Limoges, France, 51
Little, Chaplain F. K., 168

INDEX

Logan, Robert Fulton, 109, 113
London Fellowship of Medicine, xii
London University, 130
Longley, Colonel F. F., 131, 133
Lorraine Sentinel, student newspaper, University of Nancy, 153
Lough, William Henry, 34, 51
Lourdes, France, 157
Lycée de Beaune, 62
Lyon, France, 51

Maclean, George, 130, 137
McDonald, Lieutenant Colonel Otis H., 55
McKillip, Major Georg B., 55
Mankiewicz, Private First Class Herman J., 60
Marbury, Elizabeth, 149
Marne, Second Battle of, 22
Marshall, Colonel George Catlett, Jr., 1, 42, 43
Massachusetts Agricultural College, 7, 57, 82, 83, 85, 87
Massey, Lieutenant Franklin F., 149
Matthews, Captain J. D., 23
Medical College (AEF University, Beaune), 54-55
Mehun, France, 9
Memorial Day, 126, 123-33, 169
Merrian, Vera H., 66
Metcalfe, Private First Class Thomas P., 86
Meureault, France, 41
Meuse-Argonne, Battle of, 22
Michener, James, xv
Michigan Agricultural College, 86
Michigan State College, 82
Military police, 23-24
Miller, Second Lieutenant Richard C., 86
Minnesota Agricultural College, 172
Missouri National Guard, 14
Molière, Jean Baptiste Poquelin, 115
Monges, H. B., 99
Morgan, Captain Sherley W., 142
Mothers' Day, 126, 150

National Conservatory of Music, 64
National Defense Act, June 3, 1916, 6
Nevers, France, 9
Newlin, William J., 31
New York Evening Post, 60
New York Herald (Paris edition), 76, 153
New York Sun, 60
New York Tribune, 41, 60
New York University, 51
Nice, France, 123
Night School (AEF University, Beaune), 34

Nimes, France, 157
Ninetieth Division, 84
Ninety-First Division, 83
Ninety-Second Division, 15
North Carolina State College, 86
Norwich University, 15

Ohio State University, 86
One Hundred and First Infantry Regiment, 64
One Hundred and Fourteenth Infantry Regiment, 64
One Hundred and Fourth Field Artillery, 85
One Hundred and Seventy-Second Aero Squadron, 56
One Hundred and Sixty-Fourth Infantry Regiment, 23
One Hundred and Thirteenth Engineers, 64
One Hundred and Thirty-Seventh Infantry Regiment, 14, 15, 60
One Hundred and Twelfth Field Artillery, 85
One Hundred and Twenty-Third Field Artillery, 85
On Strategy II. A Critical Analysis of The Gulf War, by Harry G. Summers, Jr., 185
Orr, Louis, 113
Over There Theater League, 4, 5
Oxford University, 10, 130, 131, 133, 134, 135

Paris Peace Conference, 3, 41, 61, 153, 171
Parker, Captain Roscoe Edward, 62
Parker, Major William Hammond, 62
Parmer, Lieutenant Charles B., 147
Pasteur Institute (Paris), 54, 55
Patterson, Major Andrews H., 84, 92
Pau, France, 157
Pearce, Lieutenant Howard B., 99
Peixotto, Captain Ernest, 109, 110, 112, 113, 119
Penney, Mark E., 61
Pershing, General John J., 1, 17, 24, 29, 71, 98;
 education and, 7, 9, 18, 98, 179, 182, 186-87, 188-89;
 inspections and, 24, 73, 74;
 leave policies of, 3;
 the French and, 42, 123;
 the role of the AEF in Europe and, 42-43;
 venereal disease policies of, 36, 43;
 YMCA and, 2, 4
Pershing Stadium (Paris), 4, 114
Pershing, Warren, 74
Phare de France, 109
Philippines, 2, 14-15
Pigs Is Pigs, by Parker Butler, 72

Pont-à-Mousson, France, 15
Pont du Gard, France, 157
Popular Songs of the AEF, YMCA songbook, 5
Porter, Arthur Kingsley, 110
Post Schools, AEF, 31, 32, 33
Preston, Captain Eugene E., 54
Princeton University, 141
Progressivism, xiii, 6, 8, 185
Provence, region of France, 3
Providence School of Architecture, 99
Prun, Germany, 84
Puerto Rico, 2
Purcell, Lieutenant H. F., 91
Purdue University, 14, 15

Qu'est-ce Que C'est?, student newspaper, University of Toulouse, 150, 152, 158

Reber, Louis E., 56
Reed, Elsie, 172
Reed, Lieutenant Harry E., 106
Rees, Brigadier General Robert Irwin, 9, 19, 140, 174, 180, 189
Reeves, Colonel Ira Louis, 19, 23, 24, 42, 74, 82, 84, 99, 166, 167;
    appointed to head the AEF University, 14;
    as educator, 14-15, 17-18, 33-34, 35, 39-40, 60, 71, 171, 180, 189;
    as military commander, 18, 21;
    discipline and, 37, 38, 39, 43, 71-72, 73, 91-92;
    early career of, 14-15;
    educational philosophy of, 17-18;
    the French and, 127, 174;
    venereal disease policies of, 36;
    women students and, 35
Rhineland, 1
Rhode Island Agricultural College, 82-83
Rhodes Scholars, 131
Riviera, region of France, 3
Roberjob, M., 73
Robinson, Franklin Whitman, 63, 65
Romorantin, France, 9
Root, Elihu, 187
Rothamsted Experimental Station (Harpenden, England), 131
Royal College of Music (London), 131
Rupert, Captain Archie Keefer, 60
Rutgers University, 86

Sabin, Major Henry P., 115
Saillens, Emile, 110
Saint Aignan, France, 26, 169, 175

Saint Johns Military Academy, 22
Saint Mihiel, France, 14, 15
Saint Nazaire, France, 83, 123
Salvation Army, 27, 71
Sampigny, France, 14
Savigny, France, 40
Schmerber, First Lieutenant Louis John, 61
Schmitter, Lieutenant Colonel Ferdinand, 54
Schwab, Charlie, xii
Scott, Second Lieutenant Felix A., 32
Seavey, Captain Warren Abner, 42, 58
Second Division, 107
Sergeant, First Lieutenant Marshall W., 85
Services of Supply (SOS), 32, 77, 84, 167, 171
Seventeenth Infantry Regiment, 14
Seventh Cavalry, 23
Seventh Division, 15
Seventh Infantry Regiment, 22, 86
Shakespeare, William, 132
Shockley, Colonel M.A.W., 19
Shotwell, James T., 153
Shumway, Merline H., 172
Simpson, Captain Kenneth F., 141-42
Sixth Infantry Regiment, 60
Sixty-Fourth Infantry Regiment, 15, 64
Skinker, Dorothy Anne, 150
Small, First Lieutenant Philip L., 111
Snapp, Captain Roscoe R., 85, 92
Snow, Lieutenant Colonel William Freeman, 28, 58
Soldier Actor Division, 5
Soldier-Student, student newspaper University of Montpellier, 146, 152
Sorbonne, xii, 10, 63, 138, 139, 152, 153
Sougy, France, 9
Spaulding, Frank Ellsworth, 7, 30, 32
Sperry, First Lieutenant Marcus E., 61
Squire, Walter, 64
Stanton, Major Charles Beecher, 56
Stark, William Everett, 6
Stars and Stripes, 3, 27, 72, 76, 97, 182, 187
Stokes, Anson Phelps, 7, 8, 19, 138
Stone, Lieutenant Governor Mason S., 42
Stratford-on-Avon, 132
Sullivan, Mark, 149-50
Summers, Harry G., Jr., 185
Syracuse University, 61

Taft, Lorado, 56, 102, 110
Taft, William Howard, 133
Tenth Infantry Regiment, 22
Texas A&M College, 86
Thames, First Lieutenant Amistead E., 51
The American Club (Oxford), 133

INDEX

The Landmark, magazine, 133
"The War Baby," student newspaper, Farm School, Allerey, 92
Third Division, 22, 86, 107
Third Infantry Regiment, 22
Thirteenth Field Artillery, 86
Thirty-Fifth Division, 14, 15
Thirty-Sixth Division, 107
Thomas, Cyrus W.,110, 111
Thorndyke, A.H.,179
Three Hundred and Fifteenth Machine Bun Battalion, 64
Three Hundredand Fifty-Second Infantry Regiment, 5
Three Hundred and Fourth Infantry Regiment, 86
Three Hundred and Tenth Infantry Regiment, 84
Three Hundred and Twentieth Infantry Regiment, 85
Three Hundred and Twenty-Fourth Infantry Regiment, 86
Three Hundred and Twenty-Third Field Artillery, 64
Three Hundred andTwenty-Third Infantry Regiment, 84, 85
Titcomb, Lieutenant William C., 109, 114
Toman, Captain H. B., 23-24
Toul, France, 15
Townsend, Captain Harry, 113
Truxell, Private Vincent Earl, 64
Twelfth Cavalry, 23
Twenty-Seventh Infantry Regiment, 22

Union Theological Seminary (New York City), 42
United States Army:
    Air Service, 56, 99, 112, 114
    Candidates' School (Langres, France), 22
    Chemical Warfare Service, 58, 59
    Coast Artillery, 106
    Corps of Engineers, 99
    Divisional Schools, 9, 26, 31-32, 62, 83-84
    Educational Corps, 9, 33, 55, 56, 58, 63, 64, 66, 73, 77-78, 85, 86, 87, 114, 166, 175, 185
    Medical Corps, 58, 106, 133, 141, 149, 168
    Nurses Corps, 35
    of Occupation in Germany, 25, 27, 28, 84, 115-16, 138, 168, 172, 175
    Post Schools, 8, 26, 31, 32, 62, 83-84
    Quartermaster Corps, 58, 86, 87, 115, 174
    Regulations, 6, 8
    Signal Corps, 9, 56
    Tank Corps, 64, 106
    Veterinary Corps, 86
    War College, 187
United States Commission to Negotiate Peace, 153
United States Department of Agriculture, 83, 87, 89, 92
United States Employment Service, xiv
United States Forest Service, 89
United States Marine Corps, 60, 106, 170
United States Navy, 99
United States Weather Bureau, 89
University College (London), 135, 138
University in Khaki, xiii
University of:
    Aberdeen, 130
    Aix-Marseille, 138, 141, 155, 157
    Belfast, 130, 131
    Besançon, 138, 153
    Birmingham, 130, 132, 135
    Bordeaux, 138, 142, 152
    Bristol, 130
    Caen, 138
    California (Berkeley), 26, 102, 110
    Chicago, 60
    Cincinnati, 62
    Clermont-Ferrand, 138, 153
    Dijon, 57, 153
    Dublin, 130
    Edinburgh, xii, 130, 133, 135
    Galway, 131
    Glasgow, 130
    Grenoble, 138, 152, 153
    Illinois, 85, 109
    Indiana, 58, 60
    Liverpool, 130
    London, 130
    Lyon, 138, 140, 152
    Manchester, 130, 134
    Michigan, 25, 82
    Montpellier, 138, 140, 142, 157, 160
    Nancy, 138, 153
    Nebraska, 60
    Poitiers, xii, 138, 140, 152, 155, 156, 157, 160
    Puerto Rico, 22
    Reading, 131, 132
    Rennes, 138
    Sheffield, 130
    Toulouse, 138, 141, 142, 150, 152, 153, 155, 156, 157
    Vermont, 14

Wales, 130
Wisconsin, 85
Wyoming, 86
Sanitary Police Squad (AEF University, Beaune), 72
Service Battalion (AEF University, Beaune), 23

Vasselot, Marquet de, 110
Venereal Diseases, 36-37, 43
Verneuil, France, 9
Veterans' Benefits, xiv
Victorious, novel by Reginald Wright Kaufmann, 41
Voilà!, student newspaper, University of Bordeaux, 152
Volweider, M., 110
Vreeland, Major H. H., 142

Waco Morning News, 60
Waller, Elbert, 61
War Department, xiii, 184, 186, 188
Warren, Lloyd, 56, 99, 108, 109, 115, 119
Warren, Roy Everet, 26
Watrous, Major Livingston, 21-22
Watson, Mark S., 187
Watters, Wieford, 64
Webster, Henry Kitchell, 41
Wentworth, Captain Edward N., 57
West Point, 23, 25-26, 182
Westminster College (New Wilmington, Pennsylvania), 64
White, William Allen, 41, 61
Wilson, Sergeant Major Guy Douglas, 60
Witley, England, 10

Yale Art School, 110
Yale University, 7, 86, 141
Young Men's Christian Association (YMCA), xii, 27, 35, 66, 71, 72, 73, 76, 87, 97, 102, 104, 125, 130, 132, 138, 150, 152, 172, 173, 179, 180, 188, 189
  Cinema Department and movies in, 4;
  early employment in the United States military forces, 2, 3;
  education plans of, 7, 10;
  entertainment programs of, 4-5; founding of, 2;
  Sight-Seeing Department of, 10;
  sports in, 3-4;